A Gateway to Biomedical Research in Africa

A Gateway to Biomedical Research in Africa

Takafira Mduluza
Editor

Nova Biomedical Books
New York

For permission to use material from this book please contact us:
Telephone 631-231-7269; Fax 631-231-8175
Web Site: http://www.novapublishers.com

Library of Congress Cataloging-in-Publication Data

A gateway to biomedical research in Africa / Takafira Mduluza, editor.
p. ; cm.
Includes index.
ISBN13: 978-1-60021-444-8
ISBN10: 1-60021-444-4
1. Medicine—Research—Africa. 2. Biology—Research—Africa. 3. Clinical trials—Africa. I. Mduluza, Takafira.
[DNLM: 1. Biomedical Research—ethics—Africa. 2. Research Design—Africa. W 20.5 G259 2006]
R854.A5G38 2006
610.72—dc22 2006029572

Published by Nova Science Publishers, Inc. ; New York

Contents

Preface

Before biomedical and clinical trials began to cross international lines, recruiting patients was fairly simple. Recruiters had to research only one country's cultural and regulatory standards. But as trials go multinational and the various players in the global study community begin working together for the first time, sponsors discover new levels of complexity. Divergent regulations, cultures, and languages hamstring efforts to find a single approach to patient-outreach communications. What will work in developed country, may not be culturally acceptable in Africa, legal in South Africa, or meet ethical standards in the North or West Africa. To meet a country's ethical standards, biomedical researchers have to understand the regulatory culture first, which means they have to learn more than the law. Global standards remain a tiny speck on the horizon; hence biomedical researchers are facing an uphill battle to determine, on a country-by-country basis, the appropriate way to approach potential patients. Ethics committees and their review systems are central to the developing global regulatory climate. African Institutional Review Boards, research ethics committees may be independent groups, which may comprised of both medical and non-medical personnel. They are charged with ensuring the protection, safety, and well-being of people participating in clinical trials. Committees review and approve the study protocol, investigators, facilities, and methods, and examine all materials to be used in obtaining and documenting informed consent of study participants. This include patient-recruitment communications and above all may have limitation or no jurisdiction over determining the levels of incentives appropriate or necessary for communities for reasons unclear thereby creating disparity between studies in developed and developing countries. Compliance becomes a worldwide job and the amount of vaccines, drugs and socio-economical studies to be conducted demand huge populations whose compensation may not be feasible to most funders. Typically, global study managers have little knowledge of regulatory or other recruitment conditions in countries outside their own, where trials take place. Recognizing that global patient-recruitment efforts hinge on accurate expectations about regulatory and ethical standards, using local contacts and key opinion leaders to contact the most influential and experienced people available in the African communities maybe the most preference for patient-recruitment strategies around the world and in particular African communities.

The wildly divergent expressions from ethics committee members and African researchers as presented in this collection, present a challenge to multinational patient recruiters, but they open up some intriguing opportunities as well. African regulatory attitudes about tactics and patient outreach in clinical trials may actually be more favorable than most people, including country study managers and ethics committee members, think. The biomedical research teams can seize an opportunity to educate the clinical study community - including country study managers and ethics committees - on a broader definition of the research protocols to include building patient awareness about biomedical and the clinical trials. Most African ethics committees rely on cultural assumptions, but they are open to cultural changes. By firmly, but sensitively, presenting new patient-recruitment tactics wherever the opportunity emerges, sponsors can speed patient recruitment while keeping it culturally appropriate and within the bounds of ethical standards in remote and developing country.

Instead of assumptions of the community or groping around seeking permission to enter any African society? The best way is to become part of the society and work from within the society with the selected societal gatekeepers leading the way forward in patient recruitment while the stage on biomedical sciences would be the sole responsibility of the researchers away from the realms of the patient recruitment. Study materials should be relayed to the ethics committee and requested to assist in identifying the key specific personnel in the concerned communities. Working closely with appointed country study native manager, biomedical researchers can choose the most appropriate tactics, without caving in to assumptions and conventions. The only way to expand the boundaries is to push the boundaries, and see what gives way. As global clinical trials evolve during the next decade, they provide an opportunity for sponsors and sites to continue pushing the boundaries. Researching and respecting cultural and ethical guidelines for each country is vital. But the goal is to challenge those unreliable assumptions and unquestioned precedents that tend to limit patient recruitment. The more biomedical researchers learn/understand about African ethics committees and the community hierarchy, the more opportunity they have to leverage each country's desire to reap the rewards of bio-clinical research, which is the best motive for effecting cultural, regulatory change, health improvement and understanding the mechanisms of diseases.

The Book

The book reveals the necessary steps required by researchers from both developing and industrialized areas to conduct biomedical research in resource poor communities. The book goes beyond the ethical consideration and concentrates on issues considered minor by many researchers during planning of their research conduct. The communities considered are diverse both culturally and geographically. The communities are exposed by the authors with several years' experience working in the area on hierarchy, beliefs and fears, the economical and social aspects. Probably the education levels and how this is developing, impacting on research have been discussed. The boundaries and religion that make each community as an important aspect demarcating Africa into the social structures. The book is of value to

Biomedical Researchers involving humans, professional researchers testing drugs, equipment and vaccines from companies rearing to conduct research in resource poor communities. Social workers and including relief aid organizations, students, and teachers on aspects of research involving less developed communities. The work has been achieved by selecting prominent researchers who grew up and are involved in research work directly in the regions, with several years of community contact. Planning and preparing to undertake biomedical research in resource poor communities of Africa goes even deeper than the normally ethical consideration. Understanding the social structure for informed consent and community cooperation.

The book is of great value to University/College teachers, students and researchers planning to undertake organized trials and research in African regions/communities. The book is informative and acts as a reference by ethical committees considering proposal to work in African communities. Also this collection is of value to the sponsor/donor organizations to finance research in the selected areas, would give information of the population structure, and the likely rigorous requirement to initiate the study. The contributors to this book take pride to consolidating the general CIOMS on ethics and Helsinki document on Biomedical Research Ethics. The book consolidates the generally believed ethics to conduct biomedical research in Africa and developing communities. Protecting communities requires understanding their structures, fears and economic status. This would be the first book categorizing in a nutshell the different social and cultural hierarchy of African societies that has been in existence from generations past, current and probably a very long future to come. Considering that the African communities and structures that have been discussed and exposed have been in existence for a long time but haven't been documented properly for easy reference by groups of researchers and also is a must by teachers on social aspects and on ways to get into an organized African community getting it from an African researchers' context.

List of Contributors

Takafira Mduluza, MPhil., PhD
Biochemistry Department
University of Zimbabwe
P.O. Box MP 167, Mount Pleasant
Harare, Zimbabwe
Email: mduluza@medic.uz.ac.zw

Stephen E. D. Nsimba, PhD
Muhimbili University College of Health Sciences (MUCHS),
Department of Clinical Pharmacology,
P. O. Box 65010, Dar-es-Salaam
Tanzania, Tel: 255-22-2150302-6 Ext 214; Fax 255-22-2150465
Email: snsimba@muchs.ac.tz or nsimbastephen@yahoo.co.uk

Aceme Nyika, PhD
Centre for the AIDS Programme of Research in South Africa (CAPRISA),
Nelson R. Mandela School of Medicine
P. Bag X7 Congella 4013
University of KwaZulu-Natal
719 Umbilo Road
Durban 4001
South Africa.
Email: nyikaa@yahoo.com

Paul Ndebele, MSc
Formerly: Medical Research Council of Zimbabwe
Current: Centre for Bioethics, College of Medicine
University of Malawi
P. Bag 360, Blantyre 3
Malawi
Email: pndexas@yahoo.com

Taha El-khobhy, PhD; VASCERA
Egyptian Organization for Biological and Vaccine Production
Schistosomiasis Vaccine Development Project
51 Wezarat El Zeraa St., Agouza,
Giza, Egypt; Tel: +202 336 2639; Fax: +202 760 1746
e-mail: telkhoby@yahoo.com

Buama Juliet Vakunseh, PhD
Ndeso Sylvester Atanga, PhD
Jane Akoachere, PhD
Faculty of Science,
University of Buea,
Box 63, Buea, SW Province, Cameroon

Bourema Kouriba, PhD
Centre National de Transfusion Sanguine
BP E-344 Bamako, Mali.
Tél: 00(223)2213958
Fax: 00 (223) 221 66 08
Email: bouremakouriba@yahoo.com

Joseph Mfuso Mbengo PhD
Centre for Bioethics, College of Medicine
University of Malawi
P. Bag 360, Blantyre 3
Malawi
Email: mfutsobengo@medcol.mw

Nagla Gasmelseed PhD
Department of Molecular Biology
Institue of Nuclear Medicine, Molecular Biology and Oncology
University of Gezira, P. O. Box 20 Wad Medani- Sudan
Tel: 0024951161593 – 0024951143232
Fax: 0024951153427 - 0024951146640
Email: Nag_la@yahoo.com

Nicholas Midzi MPhil.
National Institute of Health Research
Ministry of Health,
P.O. Box CY 573, Causeway
Harare; Zimbabwe
Email: nicholasmidzi@yahoo.com

Collins Stephen Ahorlu Ph.D
Research Fellow, Noguchi Memorial Institute for Medical Research,
University of Ghana
P. O. Box LG581,
Legon, Accra, Ghana, West Africa.
Tel: +233-21-501178 (office); +233 208195705 (mobile); +233-21-502182 (fax);
E-mail: cahorlu@noguchi.mimcom.net or Collins.ahorlu@unibas.ch

Sylvester Chidi Chima, M.D., LL.M
Neuropathologist and Medico-Legal Consultant
University of Northumbria at Newcastle, School of Law
Sutherland Building, Northumberland Avenue
Newcastle-upon-Tyne NE1 8ST
United Kingdom
Email: chimas01@cs.com or chimasc@hotmail.com

James Chipeta MD, PhD
Paediatric Immunologist/Lecturer
Department of Paediatrics and Child Health
University of Zambia School of Medicine (UNZA-SOM)
P.O. Box 50110, Lusaka, Zambia
E-mail: damaseke@yahoo.com

Abbreviations

AAVP	-	African AIDS Vaccine Programme
AMANET	-	The African Malaria Network Trust
AST	-	Academy of Science and Technology
AU	-	African Union
COMREC	-	College of Medicine Research Ethics Committee
CSIR	-	Council for Scientific and Industrial Research
DA	-	District Administrator
DC/RC	-	District/Regional Commissioner
DEO	-	District Education Officer
DHMT	-	District Health Management Team
DO	-	District/Divisional Officer
DMO	-	District Medical Officer
DPO	-	District Planning Officer
DST	-	Department of Science ad Technology
EDCTP	-	EU-Developing Country Trial Platform
EHT/EHO	-	Environmental Health Technician/Officer
EU	-	European Union
NEBRA	-	Networking for Ethics on Biomedical Research in Africa
EPI -	-	Expanded Program on Immunization
FIC -	-	Fogarty International Centre
FGD	-	Focus Group Discussion
F/SMOH	-	Federal/Sate Ministry of Health
FRHD	-	Federal Reproductive Health Directorate
GCP	-	Good Clinical Practice
HC	-	Health Centre
HIS -	-	Health Information Systems
HVTN	-	HIV Vaccines Trials Network
IND	-	Investigational New Drugs
ICCPR	-	International Convention on Civil and Political Rights
IMPM	-	Institute of Medical and Plants Medicinal Research
IRB -	-	Institutional Review Boards

IRENSA	-	The International Research Ethics Network for Southern Africa
MCH/FP	-	Maternal and Child Health Family Planning
MCAZ	-	Medicine Control Authority of Zimbabwe
MINREST	-	Ministry of Technical Scientific Research
MOH	-	Ministry of Health
MRCC	-	Medical Research Coordinating Committee
MRC	-	Medical Research Council
MRCZ	-	Medical Research Council of Zimbabwe
NGO	-	Non governmental organization
NHREC	-	National Health Research Ethics Council
NHSRC	-	National Health Research Sciences Committee
NIH	-	National Institute of Health
NIMR	-	National Institute for Medical Research
NCD	-	Non communicable diseases
NMCP	-	National Malaria Control Programme
NRCM	-	National Research Council of Malaria
NRF	-	National Research Foundation
OAU	-	Organization of African Unity
OHRP	-	Office for Human Research protection
PABIN	-	The Pan African Bioethics Initiative
PDPH	-	Provincial Delegate of Public Health
PMD	-	Provincial Medical Directorate
RBM	-	Roll Back Malaria
REC	-	Research ethics committees
RH	-	Reproductive Health
SAAVI	-	South African AIDS Vaccine Initiative
SAMCC	-	South African Medicines Control Council
SANBI	-	South African National Bioinformatics Institute
SARETI	-	The South African Research Ethics Training Initiative
STC	-	Scientific and Technical Committee
US-NBAC-		US National Bioethics Advisory Commission
WB -		World Bank
WHO	-	The World Health Organization

In: A Gateway to Biomedical Research in Africa
Editor: Takafira Mduluza, pp. 1-8

ISBN: 1-60021-444-4

Chapter I

Introduction: Africa and Health

Takafira Mduluza[*]

Perspective on Africa's Health Crisis

The facts of Health and Life in Africa, are still baffling and even the countries that could be considered to be fairing well economically, have all the surprises combining the current devastation by HIV/AIDS and several of the infectious diseases. Children and pregnant women are the most vulnerable in most endemic countries and those who are immuno-compromised. The children die of diseases preventable by immunization, basic sanitation and simple community-based primary-care dehydration due to diarrhoea, measles, whooping cough, tetanus and pneumonia. Malaria kills many more and has recently increased in its scourge due to development of insecticide resistant vectors and drug resistant parasites. Women suffer increasingly high rates of deaths in childbirth and meningitis and tuberculosis is on the increase while chronic undernutrition and bad eating habits of non nutritious foods is worsening the health of many. These facts are well known by now and more suffrage is Africa torn by political mismanagement, ravaged by drought and HIV/AIDS still have other compounding health disasters. Governance and world debts are draining the scanty economical income to servicing the debt than to provide decent life for its inhabitants. The debts had infringed so much on the health delivery of several developing countries especially those in Africa. This is again compounded with severe and prolonged conflicts scattered all over the African continent. The World Bank 1981, reports that most African countries saw considerable improvements in health soon after attaining independence. War, debt, consolidation of non-democratic class interests and forms of state control over economic life plays significant roles in reversing some of these early gains. The effect of intermittent droughts drove populations to desperation and since basic needs are indivisible, health cannot

[*] Biochemistry Department, University of Zimbabwe, P.O. Box MP 167, Mount Pleasant, Harare, Zimbabwe; Email: mduluza@medic.uz.ac.zw

improve in an environment where shelter, water, food, clothing, education and even personal security are in doubt.

Approaches to Health in Africa

The World Health Organization, World Bank and UNICEF agree that the major African health challenges for the century are: malnutrition; diarrhoeal diseases and associated domestic water supply and sanitation; malaria and other vector borne and communicable diseases; HIV infection and above all health financing. The World Bank states that a quarter of the population in sub-Saharan Africa eats on average less than 80 % of their energy requirements, is infested with several diseases causing organisms. In addition to emphasizing the poor living conditions of the sahel and southern Africa, the world bank points out other conditions that undermine health delivery, examples being in most African countries in which the income distribution is particularly skewed and a large part of the population is very poor, even though some countries' national income base are strong, political instability is real instability to health delivery coupled with poor infrastructure and some with large urban populations. Several of the diseases patterns have been studied and are now wells understood on ways to combat and control them. However there is a deadly synergism between infection and malnutrition taking a great toll of Africans, not only children but also on adults. A variety of water and fly born pathogens, and tuberculosis tend to exacerbate malnutrition, often interfering with the ability to absorb nutrients. In turn malnutrition makes several diseases' impact more severe. Immunization attempts are well covered but there is need to run trials for several others in the pipeline and these being directly necessary for African populations where there are increased co-infections. Sanitation and water supply is still a major cause of concern. Surprisingly about a half of African population may not be having access to safe water. This is now in the 21st century! The goal have been missed again as usual is the case with promises. Because of the high rate of urbanization sanitation is now a major problem in most urban dwelling and the processing of the waste that is inflicting on the marginalized population that depend on open systems of water supplies in the rivers which are normally heavily polluted. Rural access to a protected water source and improved sanitation has not improved at all in most developing parts of Africa. Not surprising the incidence of water born, water vectored and water related diseases continued to increase. Malaria has been increasing in many parts of Africa where about 90% of the population live in malarious areas.

Despite efforts to control malaria, it has been increasing in many parts of the world, especially Africa. The diseases affects mostly the under five years, pregnant women and the immuno-compromised. Placental malaria is associated with low birth-weight babies and occurs in about 30% of African women living in malarious zones. Malaria parasites have become increasingly resistant to common control chemical and drugs. Among the casualties of the economic decline are massive national and multinational projects to control for river blindness, sleeping sickness, soil transmitted helminths and schistosomiasis or bilharzia. Lack of political will and long time planning on sustainability have led to collapse of several international initiated projects that contributed to the control of certain fly vector transmitted diseases, including the abandonment of zones where tsetse fly habitat is regrowing leading to the resurgence of sleeping sickness. Wars cannot be attributed, let alone regional disturbances and a lack of commitment and focus on the governments. Tuberculosis is endemic in many of Africa where it is dry and dusty even despite vaccination campaigns. Of great concern is evidence that HIV infected persons are more prone at risk of developing clinical TB and eventually spreading it. Simpler diagnostic tests are needed as well as a longer lasting vaccine and short acting drugs for treating active cases.

HIV Infections

More evidence is suggesting the virus has infected a large number of Africans – including children, women and men – although they may not show the symptoms as yet. Controversy surrounds the surveys on the reported statistics so far such that all concerns must be attended to in controlling the explosion in areas so far considered having very low levels of noticed infections. But the sub-Saharan region is the most affected in the whole world. The living conditions and access to medication warrants a lot of studies in such populations that also include potential drug trials of those that could be in the long run be affordable to the populations. It is likely that life in the rural areas is already being disrupted by AIDS mortality. The infection is remaining behind undetected in a large part of the rural population where there is also co-existence of other challenging infection from malaria, schistosomiasis and tuberculosis. The number of orphaned children is growing rapidly in most parts largely affected by AIDS mortality. Even then the burden of treating AIDS patients due to opportunistic infections weighs heavily on limited facilities and personnel already strained by financial crisis and poor living conditions. So far health in Africa suggests that unless altered, the future may condemn millions of Africans to chronic malnutrition, diseases and disability, compounded by increasing perilous conditions of work and dangerous and unhealthy urban living conditions. High maternal and child mortality may not be easy to overcome in future.

Creation of a Healthy Future in Africa

Africans must participate, be engaged and struggle in creating the conditions for a healthier future. The space is available for all this but requires collaboration and co-operation with well established organization that would assist in imparting knowledge and bring in some form of initiation to rebuilding. Strengthening and establishing Research Ethics Committees (RECs) in every corner of the continent would go a long way in making the African continent accessible to biomedical research. Collaboration on projects and programs that take an intersectoral view on health could open up a space in the economic crisis within which Africans could more freely experiment with democratic forms of defining health priorities and controlling the determinants of their health. Above all, research questions need be drafted; that may encompass: taking seriously the inclusion of the complimentary agendas of the experts and ordinary people, what would biomedical research look like? What would alternative donor activities be welcome? What role would prior staff training and committees' establishment play in enhancing research? What has to be included in the school both elementary, secondary and college curriculum on biomedical research be like? What role would the curriculum play in generating biomedical health research agendas and understanding as a base of information among the people? What kind of biomedical research is needed to help the local areas, country deal with consequences of diseases and its causative agents? Incorporate the best way in having a well informed and educated population and at community levels on biomedical research and its implications to the health in general and the socioeconomic understanding of the health needs of the communities and their environments. Some of the research questions may include the following:

i. Most of the vulnerable groups identified in Africa, what are their major complaints, needs, frustrations? How are these linked to health? What are the available local material resources and human skills? What form of organization exists among the people? What could be the obstacles to utilizing such resources, skills, organizations to improve health?
ii. What is the social-cultural hierarchy of the community; who is the decision maker in the family and at the community levels.
iii. What would happen if there are some disturbances in the economy, or a simple political occurrence like elections in the community?
iv. What is financially sustainable, what is not if the participants are to be compensated?
v. What are other activities in the area? If any, consideration should also be taken on the duration these have been running, if completed, the time period, and also considering what form of incentives where awarded to the participants?
vi. The impact of other health issues in the region or community, the potentials and priorities of an indigenous people participating in some form of biotechnology applied to major health problem in the community?
vii. What is the likelihood of some co-infections, the way these could be treated and/or excluded in the participants or the cohort?
viii. Within the research agenda could there be room for the participants to influence the research, policy agendas of the donor? How can Africans best secure space, time within the study itself to form an image of their own future and research benefit?

Biomedical Research in Africa

Africa comprises of 20 % of the world's inhabitable land and is home to approximately 682 million of the world population is a place that is so conducive to a simple livelihood of different organisms in commensalisms, mutualism or parasitism manner. The oldest continent is bedeviled with several calamities and above all due to economic winds blowing in every continent, has the worst of the suffrage on its people compounded with immense debt and health burden. Research in most parts of Africa is often financed by well-resourced, developed countries, according to the regulatory framework of the sponsor, usually sometimes inappropriate to the host population (Nature 246,(1973). The research is conducted in vulnerable host communities, whom in most cases have diverse cultural backgrounds, and also having diverse daily and life priorities (Phoenix 2002). Most research to clinical trials has been well understood by the indigenous populations and seems to be aware of the benefits thereafter. The global village and educational advancement seems to be playing a major role in educating the populations and making them understand the problems facing mankind and the need to carry out further research in a way to understand diseases mechanisms and/or be able to produce vaccines to prevent further suffering (Pang 2003). Above all, the areas previously considered to be less developed seem to have well informed personnel holding key positions in educating and informing the generally required populations for the studies. Africa has in the past decade made major strides in having key educated people in Medical schools or Colleges of health sciences. This comes from the fact that in each developing country there happen to have a health training college or medical facilities that is at least staffed with members trained from the developed countries. Currently there could be several economic problems affecting these developing countries, and could be experiencing massive exodus of trained personnel but systems are still running with enough personnel to participate in research that involve local participants. These personnel play a major role in utilizing the populations and the environment while the developed countries with all resources would need to use them in strengthening the route to access the communities and with fulfilling aspects of proper information delivery.

Taking account of the least development in every country in Africa, we understand that each has a viable University with staff that is producing world class graduates that compete in several international advancement courses. The personnel, not only should we consider them to be less educated or probably, if ever there exists ethics committees, may lack capacity in reviewing research proposals, but the personnel may lack remuneration or being over burdened with other duties that the normally honorary functioning institutional review boards may not be of any priority resulting in blending these committees as understaffed and also not well versed with protocol approvals. The IRBs may be operating in complex environment where there could be power inequalities among funders, researchers, and also resources. Most IRBs in Africa seem to be independent but may lack resources support to functioning and normally have no fulltime personnel to run the affairs of such a challenging task (Peppin 1994). For example in instituting further protocol follow-up, some IRBs in Africa may not even have the vehicle to make these and observe the required adherence to protocols. Normally, who will fund such activities, since the few well trained personnel who does the initial protocol evaluation on weather this meets the ethics of the country and other priorities do not have the time and resources to undertake such a huge demanding task for free, which is mostly uncommon in developed countries. Taking this fact, where most IRBs in developing countries in their Universities and any research bodies have well paid protocol reviewers and those who follow-up the protocols; it is saddening to find out that the same does not exist in developing countries like Africa. But who is to fund such activities? The final committees that look at the protocols normally may receive a sitting allowance but the reviewers who comment on the protocols just do it for free, leaving only dedicated people to look at the valuable documents. So the local expertise and financial resources are often constrained with very few willing to offer their services on committees, which could lead to delay in approvals and further follow-up. These challenges impact heavily on the performance of country IRBs.

Host communities may often be vulnerable due to poverty, illiteracy and the burden of the diseases in the community (Phanuphak 1998; Pletscher 1993). However, much hope is lodged with the members whom are literate and look at the protocols. For the success of trials and any study to succeed, require meaningful community participation, to institute fair selection criteria of volunteers, sound and sensitive consent processes and to guard against any social harms and also to fulfill any promised obligation that may include post trial benefits and capacity development. However due to the overwhelming diseases now posing a universal challenge to global health, there have been several developments over the past decade aimed at increasing capacity for ethical review of health research in several developing countries. Theses are:

i. The World Health Organization (WHO).
ii. WHO-UNAIDS Africa AIDS Vaccine Programme;
iii. National Institute of Health's (NIH) Fogarty International Centre
iv. The NIH Department of Clinical Bioethics Forum
v. The Wellcome Trust
vi. The European Commission Global Bioethics Form
vii. The South African Research Ethics Training Initiative (SARETI)
viii. The International Research Ethics Network for Southern Africa (IRENSA)
ix. The Pan African Bioethics Initiative (PABIN)
x. The African Malaria Network Trust (AMANET)
xi. European Union (EU) Networking for Ethics on Biomedical Research in Africa - NEBRA

Africa still is in need for personnel training and development of IRBs regardless of having initiated such advancement in biomedical research. The initiative to date reveals the need to develop appropriate local ethical guidelines and policy to train REC members and to increase REC independence, diversity of membership and monitoring of approved protocols. The indigenous REC members would deal with the cultural appropriateness of studies, relevance of the research to host country and the need for local language consent forms. Literally be more appropriately trained in the evaluation of the protection of the research participants. This being the cardinal rational of the REC members but in general the reviewers of the protocols who would recommend to the REC, Africa is well equipped with the personnel who understand the required social, biomedical or even immunology concepts. However, the African reviewers to be well conversant with the clinical trials terminology, lengthy protocols, monitoring protocols and compounded with inadequate funding and ethics capacity training make review process difficulty and untimely for the researchers. Training in scientific aspects in the general members of REC is a requirement, however the researchers need to present their protocols in such a clear and understandable manner that is comprehended by the reviewers in concerned country who would elaborate in simple terminology for the REC members that comprise of non scientific members. If the REC members fail to comprehend scientific issue then the aspect should be considered by the reviewers who are generally the expert from the local area in that field who would consider if the research lacks scientific validity is de facto unethical insofar it exposes participants to risk and inconveniences for no purpose. Therefore the RECs should receive simplified version of the reviewers' comments that the research protocols are scientifically valid either via their own review or other form body. In as so far this terminology is understood, the scientific aspect is a broad, unrefined category that could include other aspects of the biomedical jargon. The composition of the RECs obviously includes some members who would understand this terminology due to their scientific background. As a result, the RECs members require continuous refresher courses to widen the scientific understanding; this faces major challenges since most IRBs lack financial funding to undertake such country specific training (Pickering 2003; Pickworth 2000).

How would this be tackled? Biomedical research protocols seeking approval should then literally include RECs training in their budget to run a discussion meeting of the terminology on the aspect to be conducted. The implementation requires further investigation but this could be the only way out if the general understanding on the subject matter is not understood by the RECs members; then the incoming protocols should be tasked to clarify its own protocols by means of sponsoring a few days' meeting to discuss the issues in broad, before its protocols could be on the agenda in other words act as prior sensitization of the protocols. However, should the investigators be involved? This leaves the local RECs in full charge in organizing their own understanding of the protocols but the researchers providing materials to be added onto the discussion since they would be paying for the meeting. It is not the obligation of the RECs or the governments concerned to nature the advancement of the member REC but the duty or obligation of the researchers and funding agencies to see that the locations where the funds are channeled have well trained members with scientific background. It is still necessary to explore in greater detail those scientific aspects that challenge RECs and design future workshops in relevant countries. Currently there is a visible development in research agendas where there is either marked concentration in funding HIV/AIDS research, Malaria vaccine testing/clinical trials, drug clinical trails as a result efforts should be concentrated in developing or making aware of the current issues by training RECs members on specific aspects of the research agenda on work to be performed in the country or region, a commendable effort that AMANET has been doing in the past half a decade. The training empowers the RECs members to make complex risk-benefit determination on the protocols. The training could be in generic approach in which this would cover general issues top of the research agenda and various global sponsors of research ethics training in Africa should co-ordinate their efforts to ensure that training suits the real needs and reaches appropriate individuals. There could be a need to have a roster for African RECs members to access and benefit the training through co-coordinated sponsor funding so that every country and almost all members would benefit.

References

New rules for biomedical research projects. *Nature 246:* 239-41, 1973

Pang, T.: Setting global health research priorities: ethics should also guide global health research. *Bmj 326:* 1399, 2003

Peppin, P.: Power and disadvantage in medical relationships. *Tex. J. Women Law 3:* 221-63, 1994

Phanuphak, P.: Trials in developing countries. *Asian Pac. J. Allergy Immunol. 16:* 137-9, 1998

Phoenix, J. A.: Ethical considerations of research involving minorities, the poorly educated and/or low-income populations. *Neurotoxicol. Teratol. 24:* 475-6, 2002

Pickering, N.: Indigenous peoples and medical research. *N. Z. Bioeth. J. 4:* 2, 2003

Pickworth, E.: Should local research ethics committees monitor research they have approved? *J. Med. Ethics. 26:* 330-3, 2000

Pletscher, A.: Research and society: a relationship with two faces. *J. Recept. Res. 13:* 1-18, 1993

In: A Gateway to Biomedical Research in Africa
Editor: Takafira Mduluza, pp. 9-18
ISBN: 1-60021-444-4

Chapter II

Conducting Biomedical Research in Africa - Important Ethical Issues That Researchers Need to Consider

Paul Ndebele*

Introduction

As biomedical research in Africa has been on the increase during the past two decades, the scientific community has found itself coming face to face with a growing number of ethical questions about its obligations when conducting research in Africa with its serious burden of disease and poverty. Africa appears to be quite an attractive place to scientists because of the advantages it presents in terms of the unlimited numbers of potential research participants, yet it presents so many challenges when it comes to the application of the ethical principles of respect for persons, beneficence, non-malefisence and justice. What is now immerging is the fact that there are many different issues that must be understood by researchers conducting research in Africa, both about the societies and about ethical questions and dilemmas. This chapter highlights important ethical issues that researchers conducting or intending to conduct research in Africa need to take cognizance of. The chapter also suggests alternative views and the way forward as a way of promoting the ethical conduct of biomedical research in Africa.

Recent years have seen growing international debate about the ethics of conducting biomedical research in Africa and other developing countries. Concerns have been and continue to be expressed from various groups about the validity of applying ethical principles and international guidelines that may be unquestionably accepted in a particular cultural setting to another cultural setting. The debate about ethics in biomedical research in Africa are taking place against a backdrop of international guidelines that have primarily been

* Medical Research Council of Zimbabwe, Harare, Zimbabwe; Email: pndexas@yahoo.com

drafted in response to the way that medical research is conducted in developed countries and, in places, may therefore be of limited relevance to developing countries. Such concerns could partly have been fuelled by fears of medical-ethical imperialism and fears about the use of double standards especially in international collaborative research. The current discussion surrounding the conduct of research in Africa is understandable bearing in mind the involvement of unsuspecting and uninformed human beings in very risky studies in the past as well as the recent experiences relating to the HIV/AIDS trials conducted in Africa. From the debates that are currently going on, it seems that most people agree on principles at the most abstract level, but usually have some differences on how these principles ought to be operationalized in any specific context.

Economies of African Countries

Populations in most African countries share several characteristics which render them potentially vulnerable to exploitation in the conduct of research and these include:

- the low levels of economic capacity;
- Limited experience with or understanding of scientific research.
- limited local infrastructure, personnel, and technical capacity for providing health care and treatment options;
- limited experience and capacity for conducting ethical and scientific review; and
- Factors that make the obtaining of individual informed consent challenging.

It is important to note, moreover, that these same concerns are not limited to Africa and other developing countries but also obtain in the context of research conducted in developed countries –although to a limited extent. As a word of caution - most of the research ethics literature on Africa leaves one with the impression that there is something called an "African culture" and that societies in Africa have been stagnant over time. I wish to acknowledge that there is so much diversity in Africa in terms of culture, urbanization, education and industrialization, and that even with all that diversity the above characteristics still hold and our views are therefore still generalizable.

Relevance of Research to African Countries

One aspect of biomedical research that has proved particularly controversial in Africa is the relevance of the research conducted to the research participants as well as their communities and nations. At present, many African countries have limited capacity to determine national health and research priorities, or to conduct relevant research. As a result, they become dependent on the interests of external sponsors. The United States National Bioethics Advisory Commission (2001) noted that developing countries in such circumstances, may find it difficult to refuse offers to sponsor research, even if it is unlikely to benefit their populations, because of the accompanying incentives, such as improved healthcare facilities and other benefits to individual researchers, their institutions and even policy makers and politicians.

Typically, developed countries sponsor or conduct research in developing countries for some combination of the following four reasons; the host country might desire information about effective and affordable interventions for an indigenous health problem, in order to be marketed in some developing countries, drugs and biological, even if already tested and approved in other countries, must be approved by national regulatory authorities, it is more efficient to conduct research in a country in which the condition being studied is more prevalent or it might be less expensive and faster to conduct research in developing countries. Conducting research in Africa often poses special challenges and these mainly arise from the combined effects of distinctive histories, cultures, politics, judicial systems, and economic situations. There is extreme poverty that afflicts communities and individuals, and primary health care services are generally inadequate, due to many factors and the high prices of drugs often places them out of reach of both individuals and developing country governments. Some justification is therefore necessary for conducting research in African countries other than its less stringent or less complex regulatory or ethical requirements. Whatever the reason or combination of reasons, sponsors and researchers must ensure that these activities are conducted ethically and that they are responsive to the health needs of the country or community and that they do not exploit either the participants or the populations of the host country

Informed Consent

To date, debates are still raging amongst researchers; ethics review committees and others regarding the procedural aspects of informed consent. The requirement to obtain voluntary informed consent from human participants before they are enrolled in research is a fundamental tenet of research ethics and was one of the main requirements proclaimed in the Nuremberg Code (1947). It has appeared in all subsequent published national and international codes, regulations, and guidelines pertaining to research ethics, including those in developing countries, such as India, Thailand, and Uganda. Nevertheless, there is an ongoing discussion about the value and importance of particular approaches to informed consent in other countries.

The Concept of the Person

Problems involving the interpretation and application of the requirement to obtain voluntary informed consent and its underlying ethical principles, arise for researchers, ethics review committees, and others. For example, the CIOMS Guidelines (1993) specifically address the practical difficulties in dealing with informed consent. The guidelines take cognizance of the fact that there are some individuals who may be relatively incapable of informed consent because they are illiterate, are unfamiliar with the concepts of medicine held by the investigators, or are living in communities in which the procedures typical of informed-consent discussions are unfamiliar or alien to the ethos of the community.

What has generally emerged from these debates is the fact that some of the interpretations and methods used in some western countries in obtaining informed consent might not be appropriate in Africa because of different cultural or social norms. According to Tangwa (2000), the problem that starts to emerge though even at this very general and higher level is the overemphasis of the U.S and international guidelines on the individual as a person who is rational and exists in isolation from others. This brings a unique challenge when brought into the African situation where a person is a part of a social group and his role in the group is defined relative to other group members. The Bantu communities of Southern Africa have the concept of *ubuntuism* and *ubuntuism* is about people and about connectedness. In isiZulu, isiXhosa and isiNdebele they say *umuntu ngu muntu ngabantu* meaning a person is a person with and amongst other people. A person in African cultures is closely related to others, such as the extended family and is also related to those who have passed (the ancestors) and those yet to be born (the future). This leads to the involvement of others in the informed consent process.

A number of procedural issues usually arise during the process of obtaining informed consent and these require careful consideration before determining whether voluntary informed consent can be obtained. These include, for example, determining when it is necessary to obtain written consent and when oral consent should be permitted; when, if ever, the need in some cultures to obtain a community leader's or a family member's permission before seeking an individual's consent; and standards of disclosure for research participants in cultures in which people lack basic information about modern science or have different explanations of disease causation which emanates from traditional and nonscientific beliefs. It is important to distinguish procedural challenges from those that reflect substantive differences in ethical standards.

Developed country regulations and international guidelines are often criticized for focusing primarily on the informed consent document and requiring written consent. This may be difficult to implement among illiterate individuals in some African villages and such documents may be viewed as dangerous to sign in countries which have experienced oppressive regimes. In Zimbabwean rural areas for example, a signed document is rather too formal and scientists would rather do with verbal consent which is also a sign that most communities trust the scientists very much and would rather even trust them with their own lives. Informed consent needs to be focused on the process, rather than on documentation. The local ethics review committees as well as the local researchers need to be given flexibility and local discretion over the documentation of informed consent in accordance with local cultural norms. Ethics review committees should grant such waivers only if the research protocols specify how the researchers and others could verify that research participants have given their voluntary informed consent

Belief Systems and Disclosure of Information

Belief systems in some African cultures do not explain health care interventions with the same concepts that are used in modern medical science. In a study conducted in Gokwe in Zimbabwe by the author and others, questions were asked to household heads in villages about malaria and the household heads were talking about "nyongo" which they said was caused by eating immature sugar cane and water melons. The "nyongo" had the same signs and symptoms as malaria and to them mosquitoes were not a serious problem besides the occasional bites. The striking thing was that there were many episodes of this "nyongo" during the rainy season, a time during which the community also has a sudden rise in malaria cases. The investigators had to go back and re-do their questionnaires in such a way that they could now capture the episodes of "Nyongo" or malaria. Some scientists would be against such an idea as it is seen as reinforcing wrong information, but for the team it was the only way they could capture the information they needed (but at the same time they made sure to provide information on malaria). Researchers, in consultation with community representatives need to develop procedures to ensure that potential participants understand the information provided in the consent process and this entails the use of community advisory boards or such similar structures. Researchers and ethics committees also need to assess trial participants understanding of the study itself as well as study procedures.

Involvement of Community Leaders in Informed Consent

One issue that continues to attract debate in the area of international research is that of community leaders and their role in facilitating access to the community. In rural Africa, investigators often need to obtain permission from a community leader or village council before approaching potential research participants. Where custom requires that permission of a community representative be granted before researchers may approach potential research participants, researchers need to be sensitive to such local requirements and in no case may permission from a community representative or council replace the requirement of a competent individual's voluntary informed consent. The role of community leaders or elders needs to be viewed as an integral part of the process of recruiting research participants. The community leaders otherwise need to be taken as gatekeepers and as such as another layer of protection besides the ethics review committee as they assist in ensuring that their people are not abused by unscrupulous researchers from the cities. One of the most important precautionary measures that need to be considered when dealing with gatekeepers is that permission obtained from the gatekeeper must not be substituted for the need to take separate and full informed consent of the participants.

For obtaining permission of the gatekeeper, no pre-condition, demanding, sharing of information or data obtained should be accepted. In the process of research or data collection, adequate care should be taken to ensure that the relationship between the gatekeeper and the participants is not jeopardized in any way. Greater care should also be exercised in protecting participants from possible victimization by gatekeepers while publishing and disseminating results of research. Acknowledging the existence of gatekeepers is to acknowledge the existence of other political systems and structures in rural African societies besides the western democratic model of governance which hinges on individual freedoms.

Involvement of Family Members in Informed Consent

Another common feature in African society is that of potential participants consulting other persons such as family members during the informed consent process. The involvement of family members needs not be discouraged since involving family or community members in the informed consent process might even enhance the individual's ability to make his or her own choices and to give informed consent or informed refusal. When a potential participant wishes to involve family members, researchers should take steps to accommodate this request. Researchers need to acknowledge the close-knitness in African societies and the interdependence that exists among family members. Our view is that this reliance on other members of the family also needs to be seen again as an added layer of protection. The paternalistic view which is adopted by many guidelines also applies to rural African societies which have put up some measures to protect 'those who cannot make decisions for themselves'.

Researchers need to note that upholding the principle of respect for autonomy does not only entail respecting the individual person, but should even go as far as respecting their convictions, culture and traditions. The above has also illustrated that informed consent is a complex process involving negotiation between the researcher and the research participants (and their significant others) and that this process varies from individual to individual and from society to society. The discussion suggests the need for researchers to be trained in skills to make sensitive judgments according to the demands of the situation.

Right to Withdraw

In the informed consent process, the right to withdraw is very important and serves to empower the research participant. In Africa, unfortunately, the right to withdraw may mean very little or nothing to the research participants. The right to withdraw may mean nothing when people can only access treatment through participating in research. It also means very little if withdrawal can affect the relationship of the research participant with the only physician serving that area. The right to withdraw, especially in studies that provide anti retroviral treatment for HIV/AIDS, means nothing if results in the withdrawal of life long treatment. The right to withdraw may mean little if withdrawal results in drug resistance and researchers take all steps necessary to ensure that drug resistance does not develop. The fact that community members trust of researchers means that they may forfeit their right to withdraw from research as they believe that withdrawal does not always serve their interest.

Therapeutic Misconception

Several studies conducted in Africa to assess informed consent, have shown that the issue of therapeutic misconception requires some serious attention in biomedical research in Africa. Therapeutic misconception is a belief by study participants that they are participating in research for the benefit of treatment. Studies conducted among microbicide and vaccine trials, participants revealed that some participants developed some sense of false confidence and began to feel that they could engage in risky behaviours since they thought they were being protected by the microbicide or vaccine (*Ramjee, Morar, Alary, Mukenge-Tshibaka, Vuylstek and Ettiègne-Traoré, et al, 2000)*. Researchers in Africa have to deal with this serious challenge and come up with ways of disclosing information that ensures that participants are well informed about research and its implications and hence take their own informed decisions with regards to what happens to their bodies. Such innovative processes should be flexible enough to acknowledge the differences in understanding among individuals. Researchers need to strive for a balance that maximizes potential benefits and opportunities, recognizes and extends individual autonomy, and minimizes risks associated with participation in biomedical research.

Undue Inducements and Coercion

The issues of undue inducements and coercion also take centre stage in Africa especially in the context of international research. "The scientist" in the African context is "someone who knows all and as someone who should not be questioned". The problem of coercion becomes even more serious when a physician is involved in the recruitment of patients or members from the community served (*Nuffield Council on Bioethics, 2002)*. The problem of undue inducements is created by the poor economies of developing countries and people are forced to consider the benefits and ignore the risks associated with a particular study. In some cases, the amount offered to participants as compensation appear reasonable to the sponsors and other participants in developed countries, and yet they are irresistible in developing countries. Our view is that in the interest of protecting locally funded research, some of the benefits may have to be converted into community benefits such as supporting clinics or engaging in other community projects which promote good health. At least that way we can ensure that we are dealing with the 10/90 gap in an ethical manner.

Standard of Care Issues

Perhaps the most contentious issue at present is the standard of care that should be provided to participants in the control group of clinical trials. A few countries in Africa such as Uganda and South Africa, have formulated their own national ethical statements, which provide an intermediate step between the principles set out in international guidance and decisions about how these should be applied within these countries. However, national and international guidelines cannot be effective unless countries have appropriate mechanisms for their implementation. Another question that is receiving an increasing amount of attention - particularly within the context HIV/AIDS research - is what should happen once a research project in a developing country is completed. Concerns include:

- Ensuring that the intervention if found successful can be availed to the whole community (and or country).
- Ensuring that researchers receive appropriate recognition of, and where appropriate a share of the financial rewards for, their endeavors;
- How to sustain improvements to healthcare systems made during research; and
- In the vaccine and microbicide arena what happens to those who are infected during the trial.

The questions have been extending beyond the research participants to include the community at large. For some, the answer is that everyone should be given the most effective treatment, regardless of cost. Many developing countries are wrestling with issues of poverty, basic health care, and sanitation, and are unable to provide every sick person with cutting-edge or even effective treatments. The Helsinki Declaration's requirement that every patient be given the best possible treatment that would otherwise be available, worries proponents of egalitarian access to care because of its implication that if no treatment is available, as is the case in many developing countries, then doing nothing for a control group would satisfy the declaration's principle and hence the question that needs to be asked; "Is no standard a standard at all?"

Over-Researched Communities

In several African countries, there are some research centers that are internationally known as centers of excellence and are situated in poor rural areas where major studies within the countries, are conducted at these centers as well as their surrounding areas. Some of these areas are now over-researched as indicated by the literature from these areas. In some situations this phenomenon has resulted in the stigmatization of the communities. There are examples of some communities in Southern Africa who have now received various labels associated with HIV since most of the studies on HIV in the countries have been conducted in those communities. The communities in these areas serve as "Human Laboratories" and end up bearing all the burdens of research. Researchers need to ask themselves the reason why they have chosen a particular area for their study. Is it because the people are too receptive or is it to take advantage of the existing infrastructure. Certainly burdens and benefits from research need to be shared out equally among all communities.

Challenges Related to Ethical Review in Africa

Most journals now insist on evidence of ethical review of studies before they can accept articles for publication. It is therefore important for researchers to ensure that they submit their proposals to well constituted and functional ethics committees for review and monitoring. This also ensures that any ethical and methodological challenges are addressed appropriately. Unfortunately a significant proportion of the institutions in Africa do not have ethics committees and some ethics committees have inadequate ethical review processes and mechanisms and their reviewers and members are poorly equipped for ethical review. The problem of ethics committee shopping is still rampant in some African countries (when a proposal is rejected by one ethics committee, the researchers re-submit it to another ethics committee). Some committees are poorly grounded and hence are not recognized by the researchers and also by the institutions and some are under-resourced even to the extent that they do not have office space and stationery. Other problems bedeviling ethics committees relate to the abuse of expedited review process, interference from outsiders, researchers, politicians etc (lack of independence), gender imbalance as well as the non involvement of lay persons. It is important for researchers intending to conduct research in African countries to learn more about research oversight in those countries. Researchers are also encouraged to take steps to build capacity in ethical oversight in those countries and institutions in which oversight is inadequate.

Conclusion

The above discussion has illustrated that biomedical research in Africa needs to be guided by additional principles over and above the four "universal principles". The additional ones include, but are not limited to the following; cultural appropriateness, relevance to local needs and sensitivity to local conditions. The above discussion is evidence that researchers must not only be aware of the requirements of the Declaration of Helsinki and the CIOMS Guidelines, but must also be willing and able to look beyond their own cultural values. In preparing for conducting research in Africa, researchers need to take time to understand about the settings in the specific countries and institutions. The country specific chapters of this book shed more light on the situations in specific African Countries. Researchers intending to conduct research in African Countries need to make use of the country specific chapters as part of research proposal preparation.

References

Benatar S. R. (2002) Reflections and recommendations on research ethics in developing Countries. *Social Science and Medicine*. 54 (2002) 1131-1141.

Council for International Organizations of the Medical Sciences. (1993). International Ethical Guidelines for Biomedical Research involving human subjects. *CIOMS*. Geneva

Ijsselmuiden, C., and Faden, R. (1992). Research and informed consent in Africa – another look. *New England Journal of Medicine*, 326, 830-834

National Bioethics Advisory Commission (US). (2001). Ethical and policy issues in international research: *Clinical trials in developing Countries.*

Nuffield Council on Bioethics (2002). The ethics of research related to healthcare in developing countries.

Ramjee, G., N. S. Morar, M. Alary, L. Mukenge-Tshibaka, B. Vuylsteke, Ettiègne-Traoré, V. Chandeying, S. S. Abdool Karim and L. van Damme, on behalf of the COL-1492 study group. (2000). Challenges in the conduct of vaginal microbicide effectiveness trials in the developing world. *AIDS* 14(16): 2553-2557.

Tangwa G. (2000) the traditional African perception of a person: Some implications for Bioethics. *Hastings Center Report* 3095):39-43.

UNAIDS. 2000. Guidance document on ethical considerations in HIV preventive vaccine research.

World Medical Assembly (2000). Declaration of Helsinki: Ethical principles for research involving human subjects. *WMA*; Geneva.

In: A Gateway to Biomedical Research in Africa
Editor: Takafira Mduluza, pp. 19-38

ISBN: 1-60021-444-4

Chapter III

Contemporary Ethical Issues and the Regulation of Biomedical Research in African Communities

*Sylvester Chidi Chima**
Neuropathologist and Medico-Legal Consultant
University of Northumbria at Newcastle, School of Law
Sutherland Building, Northumberland Avenue
Newcastle-upon-Tyne NE1 8ST; United Kingdom

Abstract

Contemporary debate on biomedical research in Africa has focused amongst several issues on the interpretation and application of international ethical codes such as the Declaration of Helsinki [1] and other guidelines [2, 3], which are designed for the ethical conduct of biomedical research in humans. Some ethicists have argued that to remain valid, ethical codes require constant reinterpretation, in light of practical experience in the field [4, 5]. International ethical codes in general have been criticized for not fully protecting vulnerable populations because of the absence of procedures for enforcement, nor penalties for breach, and for not having the force of law [4, 6]. The Helsinki Declaration in particular, has been criticized for requiring only that researchers "be aware of" other ethical, legal, and regulatory requirements, without the caveat that its guidelines should not be used to sidestep more vigorous regulations [7]. Bioethicists and legal experts have argued that because of the hazardous nature of the research enterprise, a conflict of interest may arise between the need to test hypotheses and the requirement to respect and protect research participants. This conflict may require regulatory oversight to protect and enable participants to exercise their right of self-determination. The most critical issue in exercising this freedom of choice is the disclosure of information. Due to

* Correspondence Address: 384A Green Lanes, Palmers Green; London N13 5PD; England; E-mail: chimas01@cs.com or chimasc@hotmail.com

reported abuses of international ethical codes, biomedical research in Africa would benefit from local legislation and regulations that provide guidance on the role of local research ethics committees, informed consent procedures, standards of care, and important aspects of distributive justice. The African Union (AU) should consider legislation and directives on biomedical research, similar to directives developed by the European Union (EU) [8, 9], which are binding on but adaptable to the laws of individual member states. These could simplify the regulatory and administrative provisions, and enhance the ethical conduct of biomedical research in African communities [10][1].

Introduction

Biomedical research may be defined as a class of activity designed to develop or contribute to generalizable knowledge in the form of theories, principles or relationships, or the accumulation of information on which they are based that can be corroborated by accepted scientific methods. [2] In this context research includes both medical and behavioral studies pertaining to human health, but is distinguished from the diagnostic and clinical practice of medicine, which is designed for observation, inference, and the advancement of clinical practice, public health, and other forms of healthcare, thereby contributing directly to the health of individuals and communities. Sometimes, research and practice are conducted simultaneously, such as when research is designed to obtain new information about the efficacy of a drug or other therapeutic, diagnostic, or preventive modality usually in clinical trials, including randomized clinical trials (RCT). In recent times these types of activities have been undergoing rapid expansion in African communities because of such factors as the need to achieve the United Nations Millennium Development Goals (MDGs), designed to assist amongst other things in reducing child and maternal mortality, and combating deadly infectious diseases such as Malaria, HIV/AIDS which have had a huge impact on morbidity and mortality in African communities [11]. Another reason for the recent increase in biomedical research activities in Africa and other countries of the 'South' is probably due to the fact that the global pharmaceutical industry may see these populations as a large base for testing new drugs and pharmaceutical preparations, without the restrictions imposed by the current regulatory climate and legal concerns for human rights in more developed countries, such as those imposed by current EU Directives [8,9], which are designed to provide protection for human subjects of biomedical research in European communities. These directives and regulations have provided guidance on the composition of research ethics committees (RECs), by clearly defining what a research committee is [8, 10], and also providing guidance on procedures for obtaining valid informed consent, and legislation for certain aspects of distributive justice such as requiring mandatory insurance protection for human subjects participating in clinical trials before such research activities are approved by a local research ethics committee (REC). In addition these directives make it mandatory for EU member states to adopt local legislation designed to bring these regulations into force in individual member states. This requirement has led the UK department of health to issue

[1] Chima SC. Global Medicine: Regulation of Biomedical Research in Africa. *BMJ* 2006; 332:848-51

local legislation[2] [12], which provides a mechanism for compliance with current EU Directives, and also provides detailed definitions of informed consent, as well as a hierarchy of relatives or persons who can provide proxy consent for those individuals incapable of giving consent for themselves, especially children and mentally incapacitated adults. Similarly, in the United States of America (USA), there are Federal laws as contained in CFR 45.46[3] [13] designed to regulate the conduct of biomedical research. In 1991, the US Department of Health and Human Services (DHHS) issued a 'Common Rule', listing regulations for the conduct of human biomedical research. This rule required eight elements of information about a research study that must be disclosed to the participants prior to the commencement of research or when seeking consent from a research subject or his/her representative. These elements include a statement that the study is investigational along with a description of the research and its objectives, a description of foreseeable risks and benefits to participants as well as others, information about reasonable alternatives, a statement clarifying the implications of participation for the subjects confidentiality, a statement about compensation if injury occurs, information about how the subject can obtain answers to pertinent research questions, and a statement regarding the voluntary nature of research participation and subjects right to withdraw without losing any benefits already accrued. The common rule also defines optional elements that could be required depending on circumstances such as the possibility of unforeseeable risks, circumstances under a subject may withdraw fro the study, the possible consequences of withdrawal, any additional costs to the subject, provision of information obtained during the study that may affect a subjects willingness to continue, and the total number of participants. The common rule requires that informed consent be documented in writing for virtually all research involving greater than minimal risk and delegates the primary responsibility for research oversight to institutional review boards (IRBs) [14] In light of the above developments, it is imperative for African countries through the AU to endeavor to establish similar regulations for oversight of biomedical research being conducted in African communities. While it is recognized that certain countries in Africa such as Egypt, South Africa, Nigeria, and others may have established American style IRBs and local RECs, the reality is that most of these are not effective in monitoring the research currently being conducted in local communities [15]. Moreover, the requirements for numerous IRBs as prevalent in North America and the United Kingdom, is resource intensive, both in terms of trained manpower and funding, and clearly outside the reach of most African countries at the present time. Further, the proliferation of RECs has sometimes been criticized for being idiosyncratic with a tendency to concentrate on mundane issues rather than focusing on the protection of research subjects from harm [16, 17]. Therefore the emphasis in Africa should be on developing national and regional regulations and protocols, which are responsive to the socio-cultural and economic needs of most African countries.

---

[2] The Medicines for Human Use (Clinical Trials) Regulations, Statutory Instrument No. 1031, Her Majesty's Stationery Office, London. 2004.

[3] US Code of federal regulations for ethical conduct of biomedical research. 45 CFR 46, Federal Register 1991; 56:28012

African Communities as a Potentially Vulnerable Population Group

The issues of poverty and lack of education, and debilitating health conditions such as malaria, HIV-AIDS, tuberculosis and maternal and infant mortality plague African communities [11]. According to recent research reports, over 70% of the 40 million humans with HIV-AIDS live in Africa, with at least 1 in 5 African adults infected with the disease, making sub-Saharan Africa the epicentre of the global HIV-AIDS pandemic [18]. Similar dire statistics are also seen in almost all global indices of development as reported by the United Nations and its agencies [11].[4] Because of the above situation, most communities in Africa may be considered vulnerable, where 'vulnerability' refers to a substantial incapacity to protect ones interest owing to such impediments as capacity to consent, lack of education and alternative means of obtaining healthcare (due to poverty and underdevelopment), or by being subordinate members of society in a hierarchical system, such as women and children in the complex socio-cultural structures common in the African society. Therefore when dealing with communities such as this, it is imperative that the researchers and sponsors of biomedical research take special care, not only to comply with the requirements of international ethical codes and guidelines, but also to take into consideration the underlying socio-cultural factors when dealing with such communities and individuals, especially in the areas of obtaining informed consent, standards of care, and patients rights. As noted by the US National Bioethics Advisory Commission (NBAC):

> "When the US (or any developed country) proposes to sponsor or conduct research in another country, when the same research could not be conducted ethically in the sponsoring country, the ethical concerns are more profound and the research accordingly requires more rigorous justification"[5] [10,19].

Some commentators on the ethics of biomedical research have stated that failures in the informed consent process lead to serious inequities in research, especially for the poor and the less educated who bear most of the research burden [20], and others such as Karine Morin have argued for the need to protect the rights of individuals involved in biomedical research when she said:

> "*It* is essential to recognize that society's interest in knowledge may not coincide with and individual subject's interest. The individual subject stands to gain nothing and loose everything including his or her right of self–determination. Furthermore, as long as the Courts continue to interpret the doctrine of informed consent in experimentation as it applies in the context of treatment, the uniqueness of the protection need for human research subjects will be overlooked. Failing to recognize that subjects who volunteer for the sake of advancing scientific knowledge are differently situated from patients who stand to benefit from treatment, results in an analysis that misconceives the purpose of disclosure. Beyond informing the patient as to the means of available to treat him or her,

[4] United Nations Millennium Development Goals 2000-2015. www.un.org/millenniumgoals/

[5] National Bioethics Advisory Commission: Presidential bioethics commission issues report on clinical trials research in developing countries. Bethesda: NBAC, 2001. www.bioethics.gov

a subject must become a voluntary and willing participant in an endeavor that may yield no direct benefit to him or her, or worse, that may cause harm"[6] [21, 24].

Ethical Decision-Making and Research in African Communities

The United States national commission on bioethics has suggested that researchers should develop culturally appropriate ways to disclose information that is necessary for adherence to the substantive ethical standards of informed consent [4]. Similarly United Kingdom General Medical Council (GMC) has advised that when seeking consent, physicians and researchers must find out about participants individual needs and priorities including participant's current understanding of their condition, beliefs, culture, occupation, or other factors that may have a bearing on the information they require, before seeking informed consent from patients or research subjects [22]. In international research ethics, the rules for information disclosure have been outlined in ethical codes [1, 2], and some legal scholars have argued in cases such *Halushka v University of Saskatchewan* [23] and *Grimes v Kennedy Krieger Institute (KKI)* [24], that a human subject of biomedical research is entitled to 'all material information' that may enable him or her to provide consent, where such material information is defined as any information that is important to that particular patient, based on a subjective patient standard rather than on a reasonable doctor or researchers standard. These requirements for information disclosure in biomedical research have been summarized by Applebaum and others [17] into eight basic criteria as follows:

- A statement that the study constitutes research and explanation of its purposes and expected duration of subject involvement, and a description of the procedures involved with experimental procedures identified as such.
- A description of the risks and discomforts that is reasonably foreseeable.
- A description of the possible benefits and risks to the subjects and others
- Disclosure of appropriate alternative treatments if any
- A statement discussing the extent of confidentiality of records generated
- An explanation of whether compensation or treatment will be available if injury occurs
- A note of who can be contacted with questions or reports of injuries
- A statement as to the voluntary nature of participation and the subjects right to withdraw at anytime

Information Disclosure and Understanding of the Information Disclosed

Information disclosure is an important aspect of enhancing the patient's moral autonomy or ability to choose, therefore most current legislation, regulatory instruments, and ethical codes have acknowledged the need for information disclosure to the patient or research

[6] Morin K. The standard of disclosure in human subject experimentation J Leg Med. 1998; 19(2):157-221

subject before being asked to consent to treatment or participation in medical research. The amount of information required in biomedical research has been the subject of legal debate and commentators have suggested the amount of information may be affected by such socio-cultural factors as the desire for or lack thereof, based on the needs of each individual [26]. The legal requirements for information disclosure can be surmised from the case law especially from North America such as the Canadian case of *Halushka v University of Saskatchewan* [23] where the court ruled that:

> "In order for consent to be effective, it must be informed, freely given and it is the duty of the doctor to give a fair and reasonable explanation of the proposed treatment, including the probable effect thereof and any special or unusual risks. Such being the duty owed by the physician to his patient in ordinary medical practice, the duty to inform is at least as great if not greater in the case of those engaged in medical research to persons who offer themselves as subjects for experimentation, because in the latter case...There can be no exception to the ordinary requirements of disclosure as there may be in ordinary medical practice. Researchers do not have to balance the probable effect of lack of treatment against the risk involved in treatment itself. The example of risks being hidden from a patient when it is important that he should not worry is not applicable in the field of research. Subjects of medical experimentation are entitled to a full and frank disclosure of all the facts, probabilities and opinions that a reasonable person might be expected to consider before giving consent."[7]

Similarly in Weiss v Solomon [27] the Supreme Court of Quebec concluded that in purely a experimental research program, the doctor must disclose all known risks, including those which are very rare or remote and a fortiori those whose consequences may be grave. Therefore the rule has become that the human subject of biomedical research is entitled to a full and frank disclosure of all material risks which a reasonable person in a similar patient/research subject's position, would be likely to attach significance to in deciding whether or not to forego the proposed therapy or participation in a research project. This appears to support the view expressed in Article 7 of the International Convention on Civil and Political Rights (ICCPR)[8], which states that no one shall be subjected to torture or cruel, inhuman or degrading treatment or punishment. In particular, no one shall be subjected without his free consent to medical or scientific experimentation [28]

Informed Consent and Cultural Plasticity: The Special Situation of African Communities

By definition, informed consent is an ethical doctrine that rests on the principle of respect for persons and their right to self-determination (respect for autonomy). One aspect of the recognition of the moral autonomy or rights of the research subject is the recognition of the need for adequate information, which will enable the individual to accept or refuse therapy or

[7] Halushka v University of Saskatchewan [1965] 52 WWR 608

[8] International Convention on Civil and Political Rights (ICCPR), United Nations, 1966

participation in a research project. As stated by the court in the American case of Canterbury v Spence [29]:

> "The patients right to self- determination can be effectively exercised only if the patient possesses enough information to enable intelligent choice...informed consent are a basic social policy for which exceptions are permitted. These exceptions would include intellectual capability such as unconsciousness and extreme psychological threat...True consent to what happens to one's self is the informed exercise of a choice, and that entails an opportunity to evaluate knowledgeably the options available and the risks attendant upon each...From these axiomatic considerations springs the need, and in turn the requirement, of a reasonable divulgence by the physician to the patient to make such a decision possible."[9]

Africa is complex and multicultural society with different languages, religions and belief systems which have evolved since the emergence of Homo sapiens out of Africa over 100, 000 years ago. This has been followed by post –colonial influences and ongoing westernization, as well as endemic poverty, disease, low literacy levels and poor education [11]. Therefore individuals in this environment are bound to exhibit different levels of understanding and acceptance of the elements of informed consent. For example Nigeria, which contains 1/5th of the entire population of Africa, has 200 different languages and over 400 dialects [30]. In addition there are both large populations of Muslims, Christians, as well as practitioners of animist or traditional religions, which are worshipped individually, or in conjunction to Judeo-Christian religions [30].[10] According to a WHO report, of the 46 countries in the African sub-region, 21 are Francophone, 20 Anglophone, and 4 speak Portuguese [15]. Based on these examples, the cultural complexity of African communities cannot be overstated. Therefore in a complex multicultural society such as this, to be effective informed consent must show cultural plasticity, in other words it must be adaptable to the culture and belief systems of the local population, especially in the area of language as it applies not only to the wording of consent forms, but also when disclosing information in the process of seeking consent for individuals and communities to participate in biomedical research. Therefore it is recommended that researchers should develop culturally appropriate ways to disclose information that is necessary for adherence to the substantive ethical standards of informed consent.

Language and Valid Informed Consent

The CIOMS has given the following guidelines on language and comprehension with regards to biomedical research as shown in the table below[11]:

---

[9] Canterbury v Spence 464 2d 772 (DC, 1972)

[10] Nigeria. A paper produced by DFID Health Systems Resource Centre for the UK Department for International Development. 2000

[11] Council for International Organizations of Medical Sciences: International ethical Guidelines for biomedical research involving human subjects. Geneva 2002.

Language and comprehension of information in informed consent. Adapted from International Ethical Guidelines for Biomedical Research Involving Human Subjects. The Council for International Organizations of Medical Sciences (CIOMS) [2].

Language - Informing the individual subject must not be simply a ritual recitation of the contents of a written document. Rather, the investigator must convey the information, whether orally or in writing, in language that suits the individual's level of understanding. The investigator must bear in mind that the prospective subject's ability to understand the information necessary to give informed consent depends on that individual's maturity, intelligence, education, and belief system. It depends also on the investigator's ability and willingness to communicate with patience and sensitivity.

Comprehension - The investigator must then ensure that the prospective subject has adequately understood the information. The investigator should give each one full opportunity to ask questions and should answer them honestly, promptly and completely. In some instances the investigator may administer an oral or a written test or otherwise determine whether the information has been adequately understood.

Therefore I would like to suggest that when seeking to obtain informed consent in African communities, researchers should try to ensure that consent forms and supporting documentation should include a description in language the research subject could reasonably understand of:

- The nature of the proposed research or treatment
- The anticipated results/outcomes of the proposed research
- The possible alternative treatments in the context of clinical trials
- All the possible risks, complications and anticipated benefits arising from the proposed research, including alternative treatments and including non-treatment in placebo based studies
- Compensation, if any for any injuries that may arise from the proposed research
- In the alternative, where the patient has chosen not to know, a statement that the patient elects not to be informed of the elements set forth above.
- Where a patient prefers the information to be disclosed to the family members, this must also be respected and a statement to such effect documented. However, consent or lack thereof of family members, community members or village or tribal head should not in any circumstances be substituted for individual's valid informed consent. [19]

For the above to be achievable, consent documents would need to be provided in local languages or in the alternative there should be made available on each occasion, a legitimate and acceptable neutral interpreter to communicate such information to the research subject. Where research subjects elect to involve community or family members in the decision making process, the research subject must be given the opportunity to consult with the family or community in a non-coercive manner before arriving at a decision whether to give or refuse consent.

Standard of Care in Biomedical Research

The issue of the interpretation of the 'standard of care' to be applied in biomedical research in Africa has been the subject of intense debate in recent scientific literature especially since the controversies arising from the maternal HIV/AIDS studies in African countries. [31] While some authorities such have advocated a universal standard based on the best current treatment [1], others including this author have argued for a locally determined standard of care achievable in developing countries which reduces global inequalities, rather than a best available standard which typically be only available in developed countries[10, 32]. In 2000, the Helsinki Declaration was modified to state that the, 'the benefits, risks, burdens and effectiveness of any new method should be tested against those of the best current methods [1]. Further, CIOMS guidelines recommend that new treatments should be tested against "established effective therapy" [2]. However in practice, especially in resource poor countries, the best current therapy or standard of care is rarely available. Moreover, there are no clearly determined international standards for every human disease condition in every setting, and even medical experts do disagree on the best current or established effective therapy in certain disorders. Therefore a pragmatic approach should be a standard of care which is universally available and applicable in all local settings. An example of this would be the use of oral rehydration therapy for diarrhoeal diseases in developing countries, which has been found a successful cost-effective alternative, when compared with the 'gold standard' intervention of intravenous rehydration, accepted as standard practice in developed countries. [33] It is notable that in English law, the standard of care is a legally defined standard, [34] which is then interpreted in accordance with the Bolam test, which states that a doctor is not guilty of negligence if he has acted in accordance with a practice accepted as proper by a responsible body of medical men skilled in that particular art…putting it the other way round a doctor is not negligent, if he is acting in accordance with such a practice, merely because there is a body of opinion that takes a contrary view" [12] [25]. This definition of standard of care is somewhat different from the libertarian rights based doctrines prevalent in North America and some other parts of the developed world such as Australia and New Zealand. It is arguable that in the context of Africa and other developing countries the standard to be applied should be determined by the opinion of local experts in the speciality or disease based on local socio-cultural realities, while striving for a best current or gold standard which may become available locally in the future. A practical example of this ethical dilemma would be in the routine application of MRI or CT scan in the investigation and management of head injuries. While the use of these diagnostic techniques is considered the standard of care in developed countries, and any physician not applying these techniques in the management of such clinical cases would be considered negligent. In most African countries these technologies and facilities are not readily available for routine clinical practice. Should physicians practicing in such clinical settings be considered negligent simply because they do not apply these techniques in day-to-day medical practice? The answer is obviously no, because though the responsible local physician may know that this is the accepted standard of medical care in developed countries, he or she is also aware that these

---

[12] Bolam v Friern Health Management Committee [1957] 1 WLR 634, 638

resources may not be available locally, and the practicing physician can only use or apply techniques which are immediately available. In the final analysis, the standard of care to be applied in resource poor settings such as Africa should be based on the best available local standard and the ethical, economic and socio-cultural realities, rather than a best current standard which is only available in developed and rich countries in the 'North'[10,].

Research Ethics Committees: Their Roles and Limitations

RECs have a very important role to play in modern biomedical research. This role involves ensuring compliance with statutory regulations as well as ethical and moral obligations. A REC is expected to function as an independent oversight committee by making sure that all moral and legal obligations due to human subjects are fully protected. In other words, they are supposed to act as guardians and ombudsmen, advocating for the rights of the human subject and making sure these rights are protected by reviewing and approving all research protocols before allowing that research to be conducted. All international ethical codes and regulatory statutes indicate that biomedical research should not be allowed unless it has been reviewed and approved by an independent REC. The importance of ethical review cannot be overstated since this is the only accepted mechanism to provide some protection for human subjects of biomedical research especially because of the risk of unanticipated harm in human biomedical research and the conflict of interest which may arise in the goals of the researcher and that of the research subject especially in the context of commercialized research by multinational drug development companies, [10, 24] and the increasing restrictions and regulation of medical research in developed countries forcing the unscrupulous researcher to seek human subjects for medical research in less regulated or more vulnerable populations as reported by recent reports from India [37]. In the United Kingdom these bodies are organized into LRECs which approve research conducted at a single or local NHS site or and MRECs which reviews research conducted at more than one collaborating institution. In North America and other parts of the world RECS are also know as Institutional Review Boards (IRBS). However, RECS have been criticized for being idiosyncratic [16], and producing variable responses depending on panel composition and other factors [17]. Some research ethicists have observed that there appears to be substantial variations in the care with which RECS address issues of informed consent. They also appear to concern themselves more with minor issues such as the wording of consent forms *rather than the* ethical and scientific propriety of the research project. In addition, there appears to be consensus, that IRBS in the US rarely exercise there statutory powers to monitor the status of human subjects, including subject recruitment methods, information disclosure, and informing them of the right to withdraw from research projects. [17] Similarly in Europe, the proliferation of RECs has been criticized for being slow, poorly informed, leading to arguments for a more harmonized REC framework. [16] In developing countries, the primary concern in this area has been the paucity of RECS due to lack of resources, leading to the fact that most research is not properly reviewed before approval. A recent study on the presence of national research ethics committees (RECs) in the WHO Africa Region found that 36% of responding countries did not have a standing national REC, although 80% had in place an ad hoc ethical review system for review and approval of research projects. About 64% reported the presence of standing RECs, although only 44% agreed that the RECs actually functioned as envisaged [15]. From the foregoing, it appears that RECS do not fully protect the right of human subjects and it important to have additional oversight and legal or statutory remedies to protect human subjects of biomedical research [6, 24]. In light of the above reports, and in the context of a globalized medical research enterprise and continued violations of international ethical codes by researchers when conducting biomedical research in developing countries such as the cases of Abdullahi v Pfizer [10, 35] and Zango v Pfizer [36], the Tenovovir trials in Africa [38], and the Cirixan studies in Central America [39]. It is imperative that effective ethical review be available in all countries in Africa to protect local communities and human research subjects regardless of their level of development. This has necessitated an advocacy by this author, for a system of national ethics modeled on the EU and UK method, adapted to local socio-political realities as a method of choice for African countries in this initial phase of developing an effective system of ethical review of biomedical research, because of the need for cost-effectiveness, due to the combination of scanty physical resources, trained manpower, career-structures, and funding [10].

The Mechanism for Establishing Effective Recs in Africa

While the desirability and importance of ethical review of all biomedical research conducted amongst African communities cannot be overstated, the mechanism for establishing effective RECs in Africa is still a matter for debate. The current approach in European countries based on (EU) Directive 2001/20/EC [8] on the approximation of laws, regulations and administrative provisions of all EU member states relating to the implementation of good clinical practice in the conduct of clinical trials of medicines for human use, provides a good reference point and guidance for the harmonization of regulations for the ethical review of biomedical research in African countries. The EU Directive is binding but adaptable to the laws of EU member states, and provides detailed guidance on amongst other issues, the formation of RECS by defining what an ethical review committee is. It goes on to provide guidelines on good clinical practice (GCP) during clinical trials for the protection of research subjects as well as the role of research sponsors. A further EU directive 2005/28/EC [9] lays down principles and detailed guidelines for GCP with regards to investigational medicinal products. These directives make it mandatory for EU member states to introduce laws to enforce these regulations in each country. Based on this requirement, the English department of Health (DOH), equivalent to a Federal Ministry of Health, has introduced regulations known as the Clinical Trials Regulations [12], which are designed to provide a national mechanism for conduct of clinical trials and biomedical research in England. This document starts by defining what constitutes valid informed consent and rules for obtaining it by stating:

> " A person gives informed consent to take part in a clinical trial only if his decision (a) is given freely after that person is informed of the nature, significance, implications and risks of the trial and either (i) is evidenced in writing, dated and signed, or otherwise marked by that person so as to indicate his consent or (ii) if the person is unable to sign or to mark a document so as to indicate, his consent is given orally in the presence of at least one witness and recorded in writing." [13]

The regulation goes on to specify the mechanism of obtaining consent from children and other legally incompetent individuals, including a hierarchy of person who are legally competent to give proxy consent on behalf of incompetent adults and minors. These local regulations are further vested in a central research ethics committee (COREC)[14] whose duty it is to register and approve all human subject research going on in English institutions including the National Health Service (NHS), research institutes, universities and affiliated institutions. Similarly in the United States of America there is federally determined policy for the conduct of biomedical research enshrined in the US Code of federal regulations which are designed to regulate the conduct of all federally funded biomedical research within the United States including regulations for US sponsored biomedical research in other countries. [13, 24]

Implications for African Communities

While there may be no need for a wholesale adoption of these directives and regulations for African countries. The provision of such protections for the conduct of biomedical research in the developed countries of Europe and North America is a valuable lesson for African countries and the AU on the need to regulate the conduct of biomedical research in Africa in a systematic manner, by formulating guidelines for the formation of RECS, re-interpreting a standard of care suitable for application in African communities, defining rules for informed consent, and important issues of distributive justice, such as compensation for injuries sustained during biomedical research and post-trial benefits for individuals and communities that are participants in research. Since we are advocating culturally appropriate means of obtaining informed consent, the examples and definitions from Europe and North America can be adapted to suit African countries, especially in the context of current emphasis on tackling local issues of healthcare and underdevelopment through the MDG. [11] In the context of biomedical research a unified approach to the structure and formation of RECS and ethical review of sponsored biomedical research is a valid and accomplishable goal. As noted by Cathel J in the American case of Grimes v KKI, [24] researchers should not be allowed to completely immunize themselves by reliance on consents especially when information furnished to a research subject is materially incomplete. Such legal duties may additionally be warranted because of the likely conflict of interest between the goal of the

---

[13] The Medicines for Human Use (Clinical Trials) Regulations, Statutory Instrument No. 1031, Her Majesty's Stationery Office, London. 2004.

[14] Central Office for Research Ethics Committees (COREC), http://www.corec.org.uk/

research experimenter and the health of the human subject, especially in the context of commercialized research.

Distributive Justice and Biomedical Research in Africa

The principle of distributive derives from the ethical obligation to treat every person in a just an equitable manner. It may be defined as a fair and equitable and appropriate treatment in light of what is due or owed a person. In biomedical research, the critical issue is how to equally share the burdens and benefits arising from biomedical research between the research participants, research sponsors and the society at large. In Africa and other developing countries, the issue of post-trial benefits and compensation for injuries arising during biomedical research are major concerns. Differences in the distribution of the burdens and benefits of research are justifiable only if they are based on morally relevant distinctions between persons, especially the vulnerable members of society, where vulnerability refers to a substantial incapacity to protect one's own interest owing to such impediments such as poverty, underdevelopment and lack of alternative means of obtaining healthcare as is prevalent in most African communities. While researchers may not be responsible for the unjust conditions where research is being conducted, they should refrain from worsening the unjust situation of research subjects or promoting injustice. Under conditions of scarcity or commercial competitiveness, the understanding and application of justice can come into conflict with the ethical obligations of respect for autonomy, non-malfeasance and beneficence. Researchers in African communities must guard against derogation of these moral obligations. Justice seeks o reduced inequity, so that no class of people, as defined by poverty, age, race, nationality, mental competence or incompetence or conditions of health, are to be exploited as research subjects, by assuming a disproportionate share of the research burdens and risks. Because of these moral obligations, the WMA Declaration of Helsinki states:

> 'In medical research on human subjects, considerations related to well-being of the human subject should take precedence over the interests of science and society…medical research is only justified if there is a reasonable likelihood that the population in which the research is carried out stand to benefit from the results of the research.'[15]

Similarly the NBAC has suggested that where the population of a developing country has been ought as the source of research participants…international research ethics require not merely that research risks are reasonable in relation to potential benefits, but also that they respond to the health needs of the population being studied[10, 19].

Despite these international guidelines and exhortations, the NBAC found that most research participants in developing countries were not usually informed of the research risks,

[15] World Medical Association. *Ethical principles for medical research involving human subjects. Declaration of Helsinki*. Ferney-Voltaire, France: WMA, 2000. www.wma.net/e/policy/b3.htm

benefits, design methods, or about compensation for injuries arising during biomedical research [19]. Recent reports of research malpractices in the cases of Abdullahi v Pfizer [35], and Zango v Pfizer [36]. The Tenovovir trials in Africa, [38] the Cirixan study in Guatemala [39] and controversies surrounding the clinical trials of maternal-to-child HIV-AIDS transmission and prevention in Africa [31], have shown that developing country research participants are not fully protected by reliance on international ethical guidelines, case law or even local laws in many cases [35, 36, 40]. In the light of these ethical abuses, it is imperative to find acceptable mechanisms of protecting human subjects of biomedical research so that those injured during sponsored biomedical research are adequately compensated. An acceptable mechanism could be a no-fault insurance scheme binding on research sponsors as practiced in the UK where the research sponsor is legally bound to provide proof of insurance against injuries, before such research is approved, similar to what currently obtains in the UK through an industry sponsored scheme by the Association of British Pharmaceutical Industries (ABPI) which provides compensation for participants injured during biomedical research or clinical trials of investigational new drugs (INDs). Similarly, current UK regulations and COREC guidelines require that research sponsors must provide proof of adequate insurance against injuries for human subjects before approval of such research protocols by the local REC [10]. In Africa, the introduction of similar legislation and requirements should become part of the terms of reference for local RECs and should be included in national or regional ethical codes and local regulations.

Regulation of Biomedical Research in Africa: The Way Forward

Ethical Regulation

In general ethical regulation is based on moral principles, which have been universally adjudged to have prima facie applications under normal circumstances to the moral and ethical dilemmas that confront researchers, sponsors, and human subjects of biomedical research. Some of these ethical dilemmas and quandaries have been in existence since the time of Hippocrates as evidenced by the Hippocratic Oath. The modern impetus for a review of these ethical dilemmas arose out of the atrocities perpetrated during the Second World War by Nazi researchers. The pace of ethical review has also accelerated within the last two decades with the new biotechnologies applicable to healthcare and the human condition, such as the discovery of the new technologies for human DNA analysis and the recent HIV-AIDS pandemic which all resulted in the re-examination of international ethical codes and other regulatory instruments. As new technologies and diseases have arisen, new moral and ethical dilemmas have emerged. The Nuremberg trials [3] led to the formulation of ethical principles, which were described as fundamental and universally applicable to most areas of biomedical research. The core principle identified has been the need the informed consent of research participants before involvement in any form of biomedical research. Other ethical principles include respect for autonomy, beneficence and justice. These principles have been identified as necessary foundation for the conduct of biomedical research in humans.

Legal Regulation

The Nuremberg Code has been described as a summary of the legal requirements for experimentation in humans and also as the most authoritative statement of the law of informed consent to human experimentation. It forms part of the international common law and may be applied in both civil and criminal cases, by state, federal and municipal Courts within the US and other parts of the world. [24] The code requires that the voluntary, competent and understanding consent of the research subject be obtained before participation in biomedical research or experimentation. Since US jurists developed this code at Nuremberg, the question has been asked why this was not adopted as part of the standards of care for human subjects of biomedical research. Though frequently cited in US Court cases, it has never been applied in criminal prosecutions and only applied in a handful of civil cases such as *Abdullahi v Pfizer* [35] (research conducted in contravention of Nuremberg code and other international guidelines leading to death and injury to subjects. Research sponsors denying responsibility and not providing adequate compensation). *Whitlock v Duke University* [41], (research subject injured in non-therapeutic diving experiment. Cited in support of violations of informed consent procedures); and *Grimes v KKI* [25] (researchers accused of violations of informed consent by involving children in non-therapeutic research likely to harm them without disclosing the possibility of Lead toxicity as a consequence of the research project). Where the Nuremberg code has been cited, it is usually in dissent, and because of this limited application in Court decisions involving human experimentation, it has been difficult for an international 'common law' on human experimentation to develop. In addition there has not been universal acceptance of the Nuremberg code by the international medical community since it was promulgated. [42] The Nazi experiments were considered too extreme, and despite evidence of similar experiments in the USA such as the Syphilis study in Tuskegee [43] and the testing of Hepatitis virus on mentally retarded children [44]. It appears that the utilitarian principles and medical paternalism, which were prevalent in the USA during this period, absolved these atrocities and ethical violations. Noting that the code applied only to outrageous non-therapeutic experiments conducted during the war, physician groups tended to find the Nuremberg code too 'legalistic' [42], the medical profession proceeded to develop an alternative code more relevant to therapeutic research or clinical practice coupled with experimentation. The most successful of these was the *WMA Declaration of Helsinki*, adopted in 1964, which was intended as a statement of ethical principles to provide guidance to physicians conducting research on humans. The WMA declaration has undergone 5 revisions within the past 40 years. Starting with a tone of paternalistic benevolence, which stated that, *'if at all possible consistent with patient psychology, the doctor should obtain the patients freely given consent after the patient has been given a full explanation.'*[16] Compare this to the *Nuremberg Code* which states categorically that, *'The voluntary consent of the human subject is absolutely essential...'*[17] The limitations of the Helsinki Declaration and controversies surrounding its interpretation and application, such as issues related to difference between therapeutic and non-therapeutic research [6], its requirements that researchers be only 'be aware of' local laws [7], has led to

[16] Declaration of Helsinki. Adopted by the 18th World Medical Assembly, Helsinki, Finland, June 1964

[17] The Nuremberg Code (1947) In: Mitscherlich A, Mielke F. Doctors of infamy: the story of the Nazi medical crimes. New York: Schuman, 1949: xxiii-xxv.

the formulation of more appropriate international ethical codes such as the CIOMS guidelines, described as the 'implementation in practice' of the Helsinki Declaration. The CIOMS guidelines were designed to indicate and advice on how fundamental ethical principles can be applied effectively in medical research worldwide with regards to different cultures, religions, traditions, and socioeconomic and national legal and administrative contexts. [45]

Current ethical debate on research ethics in developing countries has centered on issues of informed consent [6, 10, 19, 46], standards of care [32, 33, 34, 47], ethical review and distributive justice [6, 10, 32, 48]. Other areas of concern have been on the interpretation and application of international ethical codes such as the Declaration of Helsinki, and abuses that have occurred in this context in recent years [10]. The above evidence highlights the need for developing countries to introduce a research governance controls, based on international guidelines and taking cognizance of local cultural, medical, and legal realities. The proposed regulations could provide guidance on the formation of local RECS, procedures for obtaining valid informed consent, standards of care to be applied in clinical trials and other forms of biomedical research, and important aspects of distributive justice, such as compensation for injuries arising during research and post-trial benefits. These areas are identified for urgent action because they are the source of most controversies in sponsored research amongst vulnerable populations, as illustrated by the cases of *Abdullahi v Pfizer* [35] and *Grimes v KKI* [24]. Secondly, regulations governing these aspects of research have been introduced in developed countries, for example the recent EU Directive, [8] and US Federal Regulations. [13, 24] Thirdly, while numerous ethical dilemmas and patient's rights issues may arise in research, certain rights may have only a prima facie standing and can be overridden when in competition with equally compelling moral considerations based on ethical reasoning [10].

Conclusion

In view of the violations of international ethical codes by some researchers and sponsors of biomedical research amongst vulnerable population groups, especially in the realm of commercialized research, and in view of the vulnerability of African communities due to underdevelopment, poverty, sometimes the absence of alternative means of obtaining healthcare, lack of sophistication in the context of western medicine, and differences in belief systems and cultural practices. It would be important for African countries through the mechanism of the African Union (AU), to develop culturally appropriate methods for obtaining informed consent and regulating the conduct of biomedical research in Africa, similar to the practice in more developed countries as evidenced by current EU Directives and US Federal regulations which are designed to protect human subjects of biomedical research. This will enable Africa to provide assurance to the international scientific community, that research proposed to be conducted amongst African communities can only be conducted based on the same ethical principles and legal protections routinely available to research subjects in developed countries. Such regulations would enhance and encourage more ethical research practices in Africa, improve the health of Africans and protect human rights.

Guideline for Obtaining Informed Consent: Essential Information for Prospective Research Subjects

Adapted from *International Ethical Guidelines for Biomedical Research Involving Human Subjects*, The Council for International Organizations of Medical Sciences (CIOMS) [2]; Before requesting an individual's consent to participate in research, the investigator must provide the following information, in language or another form of communication that the individual can understand:

1. That the individual is invited to participate in research, the reasons for considering the individual suitable for the research, and that participation is voluntary;
2. That the individual is free to refuse to participate and will be free to withdraw from the research at any time without penalty or loss of benefits to which he or she would otherwise be entitled;
3. The purpose of the research, the procedures to be carried out by the investigator and the subject, and an explanation of how the research differs from routine medical care;
4. For controlled trials, an explanation of features of the research design (e.g., randomization, double-blinding), and that the subject will not be told of the assigned treatment until the study has been completed and the blind has been broken;
5. The expected duration of the individual's participation (including number and duration of visits to the research centre and the total time involved) and the possibility of early termination of the trial or of the individual's participation in it;
6. Whether money or other forms of material goods will be provided in return for the individual's participation and, if so, the kind and amount;
7. That, after the completion of the study, subjects will be informed of the findings of the research in general, and individual subjects will be informed of any finding that relates to their particular health status;
8. That subjects have the right of access to their data on demand, even if these data lack immediate clinical utility (unless the ethical review committee has approved temporary or permanent non-disclosure of data, in which case the subject should be informed of, and given, the reasons for such non-disclosure);
9. Any foreseeable risks, pain or discomfort, or inconvenience to the individual (or others) associated with participation in the research, including risks to the health or well-being of a subject's spouse or partner;
10. The direct benefits, if any, expected to result to subjects from participating in the research
11. The expected benefits of the research to the community or to society at large, or contributions to scientific knowledge;
12. Whether, when and how any products or interventions proven by the research to be safe and effective will be made available to subjects after they have completed their participation in the research, and whether they will be expected to pay for them;
13. Any currently available alternative interventions or courses of treatment;

14. The provisions that will be made to ensure respect for the privacy of subjects and for the confidentiality of records in which subjects are identified;
15. The limits, legal or other, to the investigators' ability to safeguard confidentiality, and the possible consequences of breaches of confidentiality;
16. Policy with regard to the use of results of genetic tests and familial genetic information, and the precautions in place to prevent disclosure of the results of a subject's genetic tests to immediate family relatives or to others (e.g., insurance companies or employers) without the consent of the subject;
17. The sponsors of the research, the institutional affiliation of the investigators, and the nature and sources of funding for the research;
18. The possible research uses, direct or secondary, of the subject's medical records and of biological specimens taken in the course of clinical care. Whether it is planned that biological specimens collected in the research will be destroyed at its conclusion, and, if not, details about their storage (where, how, for how long, and final disposition) and possible future use, and that subjects have the right to decide about such future use, to refuse storage, and to have the material destroyed.
19. Whether commercial products may be developed from biological specimens, and whether the participant will receive monetary or other benefits from the development of such products;
20. Whether the investigator is serving only as an investigator or as both investigator and the subject's physician;
21. The extent of the investigator's responsibility to provide medical services to the participant;
22. That treatment will be provided free of charge for specified types of research-related injury or for complications associated with the research, the nature and duration of such care, the name of the organization or individual that will provide the treatment, and whether there is any uncertainty regarding funding of such treatment.
23. In what way, and by what organization, the subject or the subject's family or dependants will be compensated for disability or death resulting from such injury (or, when indicated, that there are no plans to provide such compensation);
24. Whether or not, in the country in which the prospective subject is invited to participate in research, the right to compensation is legally guaranteed;
25. That an ethical review committee has approved or cleared the research protocol.

References

[1] World Medical Association. Ethical principles for medical research involving human subjects. Declaration of Helsinki, Ferney-Voltaire, France: WMA, 2000. www.wma.net/e/policy/b3.htm

[2] Council for International Organizations of Medical Sciences: International ethical Guidelines for biomedical research involving human subjects. Geneva 2002.

[3] The Nuremberg Code in United States v Karl Brandt, et al. US Adjutant Generals Department, Trials of War Criminals under Control Council Law No 10 (October

1946-April 1949).Volume 2. The Medical Case, Washington DC, US Government Printing Office, 1947.
[4] Levine RJ. New international ethical guidelines for research involving human subjects. *Ann. Intern. Med.* 1993; 119:339-341
[5] Benatar SR, Singer PA. A new look at international research ethics. *BMJ.* 2000; 321.824-826
[6] Plomer A. The law and ethics of medical research: International bioethics and human rights. First edition London: Cavendish; 2005.
[7] Bastian H. Gains and losses for rights of consumer and research participants. *BMJ.* 2001; 323:1417- 1421
[8] Directive 2001/20/EC of the European Parliament and of the Council. *Official Journal of the European Union,* 1 May 2001. L121/34
[9] Commission Directive 2005/28/EC of the European Parliament and of the Council, 8 April 2005. *Official Journal of the European Union,* 9 April 2005. L91/13
[10] Global Medicine: Regulation of biomedical research in Africa. *BMJ.* 2006; 332:848-851
[11] United Nations Millennium Development Goals. www.un.org/millenniumgoals/
[12] The Medicines for Human Use (Clinical Trials) Regulations, Statutory Instrument No. 1031, Her Majesty's Stationery Office, London. 2004.
[13] US Department of Health and Human Services. 45 CFR 46, US Code of federal regulations for ethical conduct of biomedical research. *Federal Register* 1991; 56:28012
[14] Del Carmen MG, Joffe S. Informed Consent for Medical Treatment and Research: A Review. *The Oncologist* 2005; 10:636-641
[15] Kiriga JM, Wambebe C, Baba-Mousa A. Status of national research bioethics committees in the WHO Africa region. *BMC Medical Ethics* 2005, 6:10
[16] Ashcroft RE, Benn PMW. Reforming Research Ethics Committees: Latest Proposals are a missed opportunity for radical review *BMJ.* 2005; 331: 587-588
[17] Applebaum PS, Lidz CW, Meisel A. Consent to research in: Informed consent: legal theory and clinical practice. Oxford University Press, Oxford 1987
[18] Alin MJ, Griswold A, Scharzald H, Herman A. Ethics and the AIDS pandemic in the developing countries. *J. of the International Association of Physicians in AIDS Care* (JIAPAC) 2003; 2(2): 81-87
[19] National Bioethics Advisory Commission. Presidential bioethics commission issues report on clinical trials research in developing countries. Bethesda: *NBAC,* 2001. www.bioethics.gov
[20] Ross LF. Children as research subjects: A proposal to revise the current federal regulations using a moral framework, *Stanford Law and Policy Review* 1997; 8:159-164
[21] Morin K. The standard of disclosure in human subject experimentation *J. Leg. Med.* 1998; 19(2):157-221
[22] Research: The Role and Responsibilities of Doctors. *General Medical Council UK.* London 2004
[23] Halushka v University of Saskatchewan [1965] 52 WWR 608

[24] The combined appeals of Ericka Grimes and others v Kennedy Krieger Institute [2001] 24-C-99-000925 and 24-C-95066067/CL 193461. www.versuslaw.com
[25] Bolam v Friern Health Management Committee [19570 1 WLR 624, 638
[26] Dein S, Thomas K. To tell or not to tell, *Europ. J. Pal. Care.* 2002; 9:5
[27] Weiss v Solomon (1989) Carswell Que 72
[28] International Convention on Civil and Political Rights, United Nations, 1966
[29] Canterbury v Spence 464 2d 772 (DC, 1972)
[30] Nigeria. A paper produced by DFID Health Systems Resource Centre for the UK Department for International Development. 2000
[31] IJsselmuiden CB. Ethics of placebo-controlled trials of zidovudine to prevent the perinatal transmission of HIV in the third world. *N. Engl. J. Med.* 1998; 338:838-841.
[32] Bhutta Z. Standard of care should reflect local conditions and not the best western standards. *BMJ.* 2004 329: 114-5
[33] Bhutta Z. Standards of care in research. *BMJ.* 2004 329:1114-5
[34] Halsbury's laws of England. 4th edition. Butterworths, London. 2003. 33:621-623
[35] Abdullahi v Pfizer Inc. No. 02-9223 US CA NY. 2d Cir. 10/08/2003.
[36] Zango v Pfizer. No. FHC/K/CS/204/2001]. Federal High Court Kano, Nigeria
[37] Drug Trials - The Dark Side: This World www.news.bbc.co.uk/1/hi/world/south_asia/4932188.stm
[38] Anon. Editorial: The trials of Tenofovir trials. *Lancet* 2005; 3655: 1111
[39] Tan-Torres Edejer T. North-South research partnerships: the ethics of carrying out research in developing countries. *BMJ* 1999; 319:438-441
[40] *Baraza R. Kenyan statutory body was unaware of study. BMJ 2001; 316:625*
[41] Whitlock v Duke University (NC 1986) 637 *Supp* 1463
[42] Beecher HK. Experimentation with Man. Springfield Illinois: Charles C Thomas 1959
[43] Jones JH. Tuskegee Syphilis study in: Bad Blood. New York Free Press 1981
[44] Goldman L. Hepatitis research on children: The Willowbrook Debate. *World Medicine* September and November 1971.
[45] Idänpään-Heikkilä JE. Secretary General of the CIOMS. Applicability of CIOMS 2002 guidelines to developing countries, presentation at Santiago de Chile, October 2003.
[46] Dein S, Bhui K. Issues concerning informed consent for medical research among non-westernized ethnic minority patients in the UK. *J. R.. Soc. Med.* 2005; 98:354–356
[47] Shapiro K, Benatar SR. HIV prevention research and global inequality: steps towards improved standards of care. *J. Med. Ethics.* 2005; 1:39-47
[48] Benatar SR, Singer PA. A new look at international research ethics. *BMJ.* 2000; 321:824-826

In: A Gateway to Biomedical Research in Africa
Editor: Takafira Mduluza, pp. 39-45
ISBN: 1-60021-444-4

Chapter IV

Perspective of an Historical Africa

Takafira Mduluza[*]

Geology and Geography

Africa second largest continent, measuring 30,244,050 sq km (11,677,240 sq mi) including adjacent islands; estimated population 743,000,000. Broad to the north (c.4,600 mi/7,400 km wide), Africa straddles the equator and stretches c.5,000 mi (8,050 km) from Cape Blanc (Tunisia) in the north to Cape Agulhas (South Africa) in the south. It is connected with Asia by the Sinai Peninsula (from which it is separated by the Suez Canal) and is bounded on the North by the Mediterranean Sea, on the West and South by the Atlantic Ocean, and on the East and South by the Indian Ocean. The largest offshore island is Madagascar; other islands include St. Helena and Ascension in the South Atlantic Ocean; São Tomé, Príncipe, Annobón, and Bioko in the Gulf of Guinea; the Cape Verde, Canary, and Madeira islands in the North Atlantic Ocean; and Mauritius, Réunion, Zanzibar, Pemba, and the Comoros and Seychelles in the Indian Ocean.

African peoples, who account for over 12% of the world's population, are distributed among 54 nations and are further distinguishable in terms of linguistic and cultural groups, which number around 1,000. The Sahara forms a great ethnic divide. North of it, mostly Arabs predominate along the coast and Berbers (including the Tuareg) and Tibbu in the interior regions. Sub-Saharan Africa is occupied by a diverse variety of peoples including, among others, the Amhara, Mossi, Fulani, Yoruba, Igbo, Kongo, Zulu, Akan, Oromo, Masai, and Hausa. Europeans are concentrated in areas with subtropical climates or tropical climates modified by altitude; in the south are persons of Dutch and British descent, and in the northwest are persons of French, Italian, and Spanish descent. Lebanese make up an important minority community throughout West Africa, as do Indians in many coastal towns of South and East Africa. There are also significant Arab populations both in East Africa and

[*] Biochemistry Department, University of Zimbabwe, P.O. Box MP 167, Mount Pleasant; Harare, Zimbabwe; Email: mduluza@medic.uz.ac.zw

more recently in West Africa. As a whole, Africa is sparsely populated; the highest densities are found in Nigeria, the Ethiopian highlands, the Nile valley, and around the Great Lakes (which include Victoria and Tanganyika). The principal cities of Africa are usually the national capitals and the major ports, and they usually contain a disproportionately large percentage of the national populations; Cairo, Lagos (Nigeria), Kinshasa (Democratic Republic of the Congo), Alexandria (Egypt), and Casablanca (Morocco) are the largest cities of Africa.

Economy

Africa's climatic zones are largely controlled by the continent's location astride the equator and its almost symmetrical extensions into the Northern and Southern Hemispheres. Thus, except where altitude exerts a moderating influence on temperature or precipitation (permanently snowcapped peaks are found near the equator); Africa may be divided into six general climatic regions. Areas near the equator and on the windward shores of South East Madagascar have a tropical rain forest climate, with heavy rain and high temperatures throughout the year. North and south of the rain forest are belts of tropical savanna climate, with high temperatures all year and a seasonal distribution of rain during the summer season. The savanna grades poleward in both hemispheres into a region of semiarid steppe (with limited summer rain) and then into the arid conditions of the extensive Sahara (north) and the Kalahari (south). Belts of semiarid steppe with limited winter rain occur on the poleward sides of the desert regions. At the northern and southern extremities of the continent are narrow belts of Mediterranean-type climate with subtropical temperatures and a concentration of rainfall mostly in the autumn and winter months.

Outline of History: Early History to 15th Century

Africa has the longest human history of any continent. African hominids date from at least 4 million years ago; agriculture, brought from SW Asia, appears to date from the 6th or 5th millennium B.C. Africa's first great civilization began in Egypt in 3400 B.C.; other ancient centers were Kush and Aksum. Phoenicians established Carthage in the 9th century B.C. and probably explored the northwestern coast as far as the Canary Islands by the 1st century B.C. Romans conquered Carthage in 146 B.C. and controlled North Africa until the 4th century A.D. Arabs began their conquest in the 7th century and, except in Ethiopia, Muslim traders extended the religion of Islam across North Africa and South across the Sahara into the great medieval kingdoms of the West Sudan. The earliest of these kingdoms, which drew their wealth and power from the control of a lucrative trans-Saharan trade in gold, salt, and slaves, was ancient Ghana, already thriving when first recorded by Arabs in the 8th century. In the 13th century, Ghana was conquered and incorporated into the kingdom of ancient Mali, famous for its gold and its wealthy capital of Timbuktu. In the late 15th century. Mali was eclipsed by the Songhai empire and lost many provinces but remained an autonomous kingdom.

There are few written accounts of the southern half of the continent before 1500, but it appears from linguistic and archaeological evidence that the older inhabitants were gradually absorbed or displaced by agricultural, iron-working peoples speaking related Bantu languages who originated from near the modern Nigeria-Cameroon border. Between the 1st century. B.C. and 1500, Bantu-speaking peoples became dominant over most of the continent South of the equator, establishing small farming villages and in places powerful kingdoms, such as Kongo, Luba, and Munhumutapa. Prior to and after 1500, pastoralists moved south until they encountered the various Bantu groups and founded the kingdom of Kitara in the 16th century. They subsequently founded the kingdoms of Bunyoro, Buganda, Rwanda, and Ankole, all of which had elaborate social structures based on a cattle-owning aristocracy.

European Domination

The period of European domination of Africa began in the 15th century with Portuguese exploration of the coasts of Africa in an attempt to establish a safe route to India and to tap the lucrative gold trade of Sudan and the east coast trade in gold, slaves, and ivory conducted for centuries by Arabs and Swahili. In 1488 Bartolomeu Dias rounded the Cape of Good Hope; in 1498 Vasco da Gama reached the east coast and, the following year, India. In the centuries that followed, coastal trading stations were established by Portugal and later by the Dutch, English, French, and other European maritime powers; under them the slave trade rapidly expanded. At the same time Ottoman Turks extended their control over North Africa and the shores of the Red Sea, and the Omani Arabs established suzerainty over the east coast as far south as Cape Delgado.

Explorations in the 18th and 19th century reported the great natural wealth of the continent while capturing the imagination of Europeans, who viewed Africa as the "Dark Continent." These were key factors in the ensuing wave of European imperialism; between 1880 and 1912 all of Africa except Liberia and Ethiopia fell under control of European powers, with the boundaries of the new colonies often bearing no relationship to the realities of geography or to the political and social organization of the indigenous population. In the northwest and west, France ultimately acquired regions that came to be known as French West Africa, French Equatorial Africa, and the French Cameroon, and established protectorates in Algeria, Morocco, and Tunisia. Other French territories were French Somaliland, French Togoland, Madagascar, and Réunion. The main group of British possessions was in East and South Africa; it included the Anglo-Egyptian, Sudan, British Somaliland, Uganda, Kenya, Tanganyika (after World War II), Zanzibar, Nyasaland, Northern and Southern Rhodesia, Bechuanaland, Basutoland, and Swaziland. Following Britain's victory in the South African War (1899–1902), its South African possessions (Transvaal, Orange Free State, Cape Colony, and Natal) became a dominion within the British Empire. Gambia, Sierra Leone, the Gold Coast, and Nigeria were British possessions on the west coast. Portugal's African empire was made up of Portuguese Guinea, Angola, and Mozambique, in addition to various enclaves and islands on the west coast. Belgium held the Belgian Congo and, after World War I, Ruanda-Urundi. The Spanish possessions in Africa were the smallest, being composed of Spanish Guinea, Spanish Sahara, Ifni, and the protectorate of Spanish Morocco. The extensive German holdings—Togoland, the Cameroons, German South-West Africa, and German East Africa - were lost after World War I and redistributed among the Allies; Italy's empire included Libya, Eritrea, and Italian Somaliland.

Movement toward Independence

The Union of South Africa was formed and became virtually self-governing in 1910; Egypt achieved a measure of sovereignty in 1922, and in 1925 Tangier, previously attached to Morocco, was made an international zone. At the end of World War II a rise in international trade spurred renewed exploitation of Africa's resources. France and Britain began campaigns to improve conditions in their African holdings, including access to education and investment in infrastructure. Africans were also able to pressure France and Britain into a degree of self-administration. Belgium and Portugal did little in the way of colonial development and sought greater control over their colonies during this period.

In the 1950s and 1960s, in the face of rising nationalism, most of the European powers granted independence to their territories. The sequence of change included independence for Libya in 1951; independence for Eritrea in federation with Ethiopia in 1952 (later absorbed by Ethiopia, Eritrea became fully independent in 1993); in 1956 independence for Morocco, Sudan, and Tunisia and the return of Tangier to Morocco; in 1957 independence for Ghana; in 1958 independence for Guinea and the return of Spanish Morocco to Morocco. In 1960 France granted independence to Cameroon, the Central African Republic, Chad, Congo (Brazzaville), Côte d'Ivoire, Dahomey (now Benin), Gabon, the Malagasy Republic (now Madagascar), Mali (briefly merged in 1959–60 with Senegal as the Sudanese Republic), Mauritania, Niger, Senegal, and Upper Volta (now Burkina Faso); also newly independent in 1960 were Congo (Kinshasa)—the former Belgian Congo—and Nigeria, Somalia, and Togo. In 1961 Sierra Leone and Tanganyika (now part of Tanzania) became independent, the Portuguese enclave of São João Baptista de Ajudá was seized by Dahomey, the British Cameroons were divided between Nigeria and Cameroon, and South Africa became a republic. In 1962 Algeria, Burundi, Rwanda, and Uganda became independent nations. Remaining British possessions after 1962 were Zanzibar, which gained independence in 1963 and joined with Tanganyika to form Tanzania in 1964; Gambia and Kenya, which became independent in 1963; Malawi (formerly Nyasaland) and Zambia (formerly Northern Rhodesia), independent in 1964; Botswana (formerly Bechuanaland) and Lesotho (formerly Basutoland), independent in 1966; and Mauritius and Swaziland, independent in 1968. In 1968 Spain granted independence to Equatorial Guinea, and in 1969 Spain returned Ifni to Morocco.

In 1974 Portuguese Guinea became independent as Guinea-Bissau, and the former Portuguese territories of Angola, Cape Verde, Mozambique, and São Tomé and Principe became independent in 1975. After Spain relinquished the Spanish Sahara (Western Sahara) to joint Moroccan-Mauritanian control in 1976, a guerrilla force undertook a struggle for independence there. Under military pressure, Mauritania yielded its sector of Western Sahara to Morocco in 1979; Morocco, for its part, built fortifications in the territory and resisted pressures for independence. The Seychelles and the Comoros became independent in 1976 from Great Britain and France, respectively, and in 1977 the former French Territory of the Afars and the Issas became independent as Djibouti. When Rhodesia (formerly Southern Rhodesia) unilaterally declared itself independent in 1965, Great Britain termed the act illegal and imposed trade sanctions against the country; after a protracted civil war, however, Rhodesia gained recognized independence in 1980 as Zimbabwe. South West Africa, which had been administered by South Africa since 1922 under an old League of Nations mandate (South Africa's continued administration of the territory was declared illegal by the International Court of Justice in 1971), won its independence in 1990 as Namibia. Great Britain retains control of the islands of St. Helena and Ascension, and Mayotte and Réunion remain French. Spain retains the Canary Islands and Ceuta and Melilla, two small exclaves on Morocco's coast.

The Postcolonial Period

In the early postcolonial period the most pressing problems facing new African states were the need for aid to develop natural resources, provide education, and improve living standards; threats of secession and military coups; and shifting alliances among the states and with outside powers. Recognizing that unity and cooperation were needed, African nations established the Organization of African Unity (OAU) in 1963 in Addis Ababa. African nations were also forced to form alliances based on the cold war politics of the Soviet Union, the United States, Cuba, and other countries in order to receive badly needed aid. This period saw the overthrow of democratic forms of government and numerous coups resulting in the installation of military regimes and single-party governments.

Beginning in the late 1960s and continuing through the mid-1970s, a severe drought desiccated the Sahel region South of the Sahara. The resulting famine, disease, and environmental destruction caused the death of thousands of people and forced the southward migration of additional hundreds of thousands to less affected areas. From 1975 through the end of the 1990s, Africa continued to experience political, social, and economic upheaval. The post independence era has also been marked by a rise in nationalist struggles. Wars in Sudan, Ethiopia, and Somalia continued, and political instability in these nations worsened. Civil war in Ethiopia resulted in the birth (1993) of a new country, Eritrea. Beginning in the 1970s, Chad fought Libyan expansionist activity with help of the French military. Relations between Chad and Libya were finally normalized in 1989.

In the late 1980s, there was a decline of Marxist influence in Angola, from where Cuban troops began to withdraw in 1989, as well as from civil war–torn Mozambique. An UN-aided peace process in Mozambique culminated in peaceful elections there in 1994. However, civil conflict continued through the 1990s in Angola, as numerous peace agreements between rebels and the government was broken.

South African blacks led an enduring struggle against white domination, with frequent confrontations (such as the Soweto uprising in 1976) leading to government repression and escalating violence. Throughout the 1980s the international community applied pressure in the form of economic sanctions in order to induce the South African government to negotiate with the African National Congress (ANC). In 1989 newly elected Prime Minister F. W. de Klerk promised democratic reforms that would phase out white minority rule, and in 1992 the legal underpinnings of apartheid were largely dismantled. Consequently, South Africa's black majority participated in the country's first fully democratic elections in 1994, which brought Nelson Mandela and the ANC to power.

Other African nations began to introduce democratic reforms in the late 1980s and early 1990s that included multiparty elections; transitions to democratically elected leadership have taken place in countries such as Mali, Zambia, Benin, and Malawi. Political instability and civil strife continued to plague several regions of the continent into the late 1990s, most notably Liberia and Sierra Leone in West Africa and Congo, Rwanda, and Burundi in the Great Lakes region. Peace treaties signed in Liberia (1997) and Sierra Leone (1999) between those countries' governments and insurgents promised some hope of stability. In Rwanda in 1994 a Hutu-led government that provoked ethnic tensions leading to the genocide of nearly one million persons was overthrown by Tutsi-led forces; by 1997 there was a growing war between the Rwandan army and Hutu guerrilla bands. Also in 1997, 30 years of dictatorial rule in Zaïre were brought to an end, and the country's name was changed to the Democratic Republic of the Congo; however, the new government was soon threatened by mutinous troops who assumed control of large areas of the country; a cease-fire was signed in 1999. Nigeria ushered in a new government in 1999 with the first democratically elected president since 1983. Several African countries made positive strides in managing market-oriented economic reform in the 1990s, most notably Ghana, Uganda, and Malawi.

In 1992–93, the worst African drought of the 20th century and numerous civil wars were the primary causes of a famine that spread across portions of sub-Saharan Africa and most severely affected the nations of Somalia and Mozambique. The scourge of AIDS has continued to pose a major health threat to many African nations, as a lack of economic resources has prevented an effective response. In 1997, it was estimated that some 21 million Africans were infected with the AIDS virus; in Botswana and Zimbabwe, one out of every four adults was infected. Ethnic tensions and political instability, along with the resulting economic disruption, remained problems in many African countries. Mindful of the OAU's relative ineffectiveness in dealing with these issues and seeking an organization with greater powers to promote African economic, social, and political integration, African leaders established the African Union, which superseded the OAU in 2002.

Economy

Most of Africa's population is rural, but, except for cash crops, such as cacao, tobacco, cotton and peanuts, agricultural production is low by world standards; Africa produces three quarters of the world's cocoa beans and about one third of its peanuts. Rare and precious minerals (including much of the world's diamonds) are abundant in the continent's ancient crystalline rocks, which are found mostly to the south and east of a line from the Gulf of Guinea to the Sinai Peninsula; extensive oil, gas, and phosphate deposits occur in sedimentary rocks to the north and west of this general line. Manufacturing is concentrated in the Republic of South Africa and in North Africa (especially Egypt and Algeria). Despite

Africa's enormous potential for hydroelectric power production, only a small percentage of it has been developed. Africa's fairly regular coastline affords few natural harbors, and the shallowness of coastal waters makes it difficult for large ships to approach the shore; deepwater ports, protected by breakwaters, have been built offshore to facilitate commerce and trade. Major fishing grounds are found over the wider sections of the continental shelf as off North West, South West, and South Africa and North West Madagascar.

Much has also been written on the global effects of environmental degradation, population growth, and economic disparities. In addition, the pace of scientific development has accelerated, with both negative and positive implications for global health. Concerns for national health transcend borders, with a need for shared human security and an enhanced role for international cooperation and development. These issues have significant bioethical implications, and thus a renewed academic focus on the ethical dimensions of public health is needed.

Future developments in science and health policy also require a firm grounding in bioethical principles. These core principles include beneficence; nonmaleficence (to do no harm); respect for persons and human dignity (autonomy); and attention to equity and social justice. According to the World Health Organization; global ethical approaches should:

(1) monitor and update ethical norms for research, as necessary;
(2) anticipate ethical implications of advances in science and technology for health;
(3) apply internationally accepted codes of ethics;
(4) ensure that agreed standards guide future work on the human genome; and
(5) ensure that quality in health systems and services is assessed and promoted.

In: A Gateway to Biomedical Research in Africa
Editor: Takafira Mduluza, pp. 47-60
ISBN: 1-60021-444-4

Chapter V

Globalization of Health: Informed Consent

Takafira Mduluza*

Developing countries are now increasingly in demand of actual population based estimates of health information whose measures require the use of populations. Collection of data requires careful assessment of ethical standards and procedures related to issues of safety, informed consent, reporting and referral policies (Phoenix 2002). Much of it is now well documented and laid out procedures are available in some of the developing countries. However, some of the protocols seem to be completely full of dishonest in participation. The mostly referred to rational is the need to add knowledge and also some form of basic participant examination to be performed. Yet the same protocol would incorporate massive reimbursement if conducted in developed country. As a result most of the researcher contributing to the articles in this book advocated for a unitary set of ethical principles to be followed throughout the world regardless of the standards of living or general income of the participants (Phipps et al. 1999; Pickering 2003). However, due to controversies over research conducted in developing countries much scrutiny has intensified on all research being conducted in these settings. All research has to consider the global human rights and medical ethical principles including fidelity, truthfulness, confidentiality, autonomy and beneficence. These have to be carefully reviewed for proper research surveillance, monitoring and evaluation. These ethical issues in population-based surveys include informed safety, consent, confidentiality and reporting findings of testing. Currently, the awareness of ethical principles have reached most researchers in every corner of the world such that training may be lacking in particular protocols of research ethics but the general principle is now well understood. Implementation of the research protocols have to be standardized and modified according to country specific cultural context. Global research world is now slowly and ultimately becoming a village (Katz 1996; Kishore 2006).

Traditionally early effort to established good relationship with indigenous people in biomedical research succeeded while current more sophisticated approach with arguably

* Biochemistry Department, University of Zimbabwe, P.O. Box MP 167, Mount Pleasant; Harare, Zimbabwe; Email: mduluza@medic.uz.ac.zw

better opportunities seem to be heading towards several controversies and likely future difficulties in mobilizing study participant in less developed countries (Pickworth and Brazier 1999). Many clinical and biomedical trials have struggled with identification of appropriate representation of community interests and with concerted campaigns by nongovernmental organizations (NGOs) several training workshops have been conducted but still facing several hurdles to accessing communities especially in Africa, despite broad interest among the indigenous communities to participate (Pinkerton et al. 2002; Pope John 2002). Several studies have succeeded in developing credible working partnerships and carried out project through to completion, while some became mired in defense of general approach to prior informed consent and had poor participation. Successes of some biomedical investigations give some lessons about the role of culture, politics, and local governance in the differing outcomes of their efforts. In particular, the role of preexisting and broadly representative indigenous governance as a key factor in determining the feasibility and integrity of prior informed consent for the individual and community participation. Research in developing countries can, carefully constructed and equitably managed, produce benefits to health, conservation, and sustainable development. A critical component of the program is the development of ethically sound partnerships among diverse organizations, including the indigenous communities involved in or significantly affected by a project. To develop prior informed consent and to negotiate access and benefit sharing agreements with the indigenous communities (Pope John 2003). The increasing interactions of indigenous societies with the Western scientific community are not only inevitable in this shrinking world but capable of yielding substantial social gains for both. The critical question becomes how to develop research projects that minimize risks to participants and maximize the benefits to indigenous societies and the global community. Ethical research (a) generates useful knowledge, (b) is scientifically rigorous, (c) chooses research subjects fairly, (d) has a favorable risk-benefit ratio, (e) is independently reviewed, (f) maintains the privacy and monitors the well-being of participating individuals, and, finally, (g) requires the prior informed voluntary consent of the participants. The regulatory requirement of freely given and informed consent to participate in research is based on ethical principles including respect for persons, human dignity, and autonomy. The aim is to ensure that research subjects understand the risks of potential harm that they may be exposed to as well as the potential benefits that may accrue to them, to others, and to medical science (CIOMS, 1993;(2002). The principal concerns relate to the possibility of exploitation of local people for scientific and commercial purposes. Exploitation is a danger in any scientific research project involving human subjects because the principal interest of the investigators - scientific knowledge - is generally not as available or important to the lay participants (Kopelman 2004; Macpherson 2004). The statement does not specifically address knowledge ownership issues but rather offers a general statement regarding informed consent procedures as they relate to institutional review boards. It points out that written consent is not always appropriate and indicates that community authorities sometimes need to be consulted before individuals are approached (Jamison 1998; Marshall 1998).

That significant economic benefits may accrue to that rare venture that leads to a commercially successful pharmaceutical drug requires thorough discussion with the participants from the onset. The distribution of economic benefits in today's commercial and intellectual-property frameworks generally undervalues the contributions of biomedical research and clinical trials and leaves indigenous societies that contribute to them particularly vulnerable to exploitation. Driven by long-standing concerns about exploitation and coupled with interest in promoting incentives for diseases knowledge and access to treatment, the sharing of benefits with participating communities and nations is one of the principal objectives of the ethically sound research protocols. A variety of more modest but often important monetary and non monetary benefits can generally be developed. Discussions of prior informed consent related to biomedical researching have tended to be legalistic and focused on national authorities. Consent has been identified as a basic requirement for access to participants and associated collection of biological materials and running clinical trials. In the context of participants it is defined as "(a) consent of the contracting party which is the biomedical material provider (b) based on full and complete information provided by the potential researcher (c) prior to consent for access being granted" are nation states. The guidelines on access and benefit-sharing have modestly elaborated upon these principles. A widely accepted definition of prior informed consent in relation to epidemiology surveillance has yet to emerge. What is clear is that, for the reasons outlined above, informed consent is especially concerned with discussion and negotiation of benefits. The biomedical program approaches the questions of prior informed consent on the basis of three principles. First, it acknowledges the above-described implications of an individual from communal origin and of individuals and biological materials; and recognizes that community interests must be respected. Second, it recognizes the possibility of harm to or exploitation of community individuals and interests. Third, prior informed consent is part of a multipurpose outreach effort that includes raising awareness of the potential global values of biomedical studies and community participation as well as negotiation of benefit sharing. Standard practice is that prior informed consent may be developed at the community level before seeking it from individuals. Finally, it is assumed from the beginning that *if prior informed consent is properly developed, indigenous peoples will be able to make intelligent decisions about the risks and benefits of participation.* Several important dimensions of the social context that make prior informed consent feasible or not in biomedical research arrangements with indigenous peoples: culture, governance, and politics (2002; Corrigan 2003; Sleeboom 2005).

Nonetheless, arriving at consensus on what constitutes the relevant "community" that can legitimately make decisions regarding biomedical research is a significant challenge. Identifying the relevant social unit with which to collaborate is challenging. Is it the village, the municipality, the clan, the entire language group, or all those who inhabit the bioregion (Patil 2004; Saha 2004). Moreover, it is unlikely to be satisfying to many stakeholders and outside commentators because of governance and political issues that may impinge on the definition of the relevant unit of "community." The legal authority to give informed consent with regard biomedical research is defined in the CIOMS at the minimal level of the nation-state. For some countries in Africa, significant authorities for health management are identified in the constitution at the province/department/state or lower level, but for the most part the implications for biomedical researches are not yet clearly articulated in regulations. Legal assignment of consent authority to indigenous communities has been slow in many countries because of preexisting disputes over different power line; political and traditional or use rights between indigenous peoples and governments. However, the convention Helsinki and CIOMS strongly encourages and most countries recognize at some level the principle that prior informed consent must be obtained from indigenous and local communities for the use of participants associated with the endemic area or trial community (Bandewar 2003; Helgesson and Johnsson 2005; Srinivasan 2001). Frequently this principle is extended to cover rights to use the associated biomedical samples with or without use of the knowledge given to the participant (Helgesson and Johnsson 2005).

Even where prior informed consent is not legally required, most scientists wishing to use communities are willing and often eager to go through culturally appropriate channels to obtain it. However, their ability to do so is greatly influenced by the degree to which the process is clearly defined and communicated to them and possible to achieve with reasonable effort. A number of both general and ethnicity-specific models for the process of obtaining prior informed consent are emerging. The principal problem for the research community is the absence of a clearly delineated governance hierarchy in many African societies that formally establishes for the outside world what level of an indigenous community or nation has the authority to give consent (Marshall 2005). Moreover, the self-defined authorities of some indigenous groups overlap and may change over time normally between traditional and political. Frequently, overlap occurs because self-defined "communities" may share language, knowledge, traditions, and land-use rights with other such communities. The temporal fluidity of some authorities may derive from strong traditions of family and village or clan autonomy or other factors such as nomadism or intermarriage. Finally, indigenous governance systems are rarely acknowledged formally in national laws. Consequently, a research scientist is rarely secure, that the consent granted by an indigenous society through the locally identified mechanism is authoritative or enduring. In addition to the broad areas of culture and governance, a third important determinant of the feasibility of obtaining prior informed consent is politics at local, national, and international levels (Resnik et al. 2005). Tensions among indigenous communities, between those communities and national governments, and among national governments in discussions make it difficult to achieve clear and reasonable dialogue on issues such as intellectual property rights and benefit sharing. Emotionally charged issues related to human rights, land tenure, the ability of poor communities to exploit the patent system, and the morality of patenting inventions derived from study of living organisms frequently inject themselves into the debate (Foster et al. 1998).

Most aims of the project are multiple but centered on biomedical and drug trials most widely used by the developed world and to a lesser extend those specific for tropical and subtropical temperate zones. Rather than work with specific hospital based, the investigators normally choose to focus on the disease patterns prevalent or most often cited by community members as greater health challenge. A substantial amount of the epidemiology information is already in the public domain in one form or another, sources are the developed world library both accessible through the electronic and print libraries. Most villages in African communities are deeply stressed by poverty, a deteriorating health delivery system and natural resource base, and extraordinary population growth. Communities are further divided by religious conflicts and divergent loyalties to the ruling and the national government. As is the case for many indigenous African societies, a tradition of general communal assembly exists for decisions of generalized importance, the participants in such assemblies are traditionally-identified and the authority of those assemblies to represent the rest of the community is well established in these communities. In contrast to the situation in some developing countries in Africa, there are no indigenous political organizations that exist continuously and are authorized to speak on behalf of communities in relation to local or national biomedical issues. Biomedical research in most cases begins with the development of an elaborate informed-consent protocol concurrent with the non-biomedical or trial portions. The early work should include training participants and health workers, translating, analyzing, and disseminating biomedical knowledge to communities, collecting biomedical specimens for research. Recognizing the national stage on which the team would work, the project members, led by the local district team, begin the informed-consent process by organizing a national forum on biomedical research to draw lessons from other projects. Then followed with an invited general-information assembly of the community members, flyers in native languages, and radio spots (Molyneux et al. 2005; Rhodes 2005; Turner 2005; Wicher and Michalek 2005; Zion 2005). In addition, recruitment of at least a few local members to be part of the research team could avoid suspicion and the trained local youth or personnel would provide comments and linkage to improve the study plans integrated into the community. The heart of the community consent process could be a one act play performed in native languages designed by the researchers and the local personnel project members and enacted by a group of local apprentices. The play would depict the aims and methods of the project, the proposed near- and long-term elements of the benefit-sharing plan, the low probability of a commercially successful scientific discovery, and the more likely concrete benefits from other parts of the project. A host-country institution in a leadership role is a must for a successful biomedical research program and for the project to satisfy one of the minimum criteria.

If indigenous communities are to find the space to negotiate on their own behalf with outsiders rather than through national governments, universities, or external NGOs, they will often need the authority that Western-style organizational and accountability systems provide. Otherwise they will continue to be susceptible to paternalistic efforts to exploit or protect their interests by the organizations they depend upon and those that identify themselves as their protectors. While an important part of the development community today, most NGOs represent special interests and are not generally accountable directly to communities. Similarly, universities often attempt to negotiate on behalf of indigenous communities with which they work and, while less prone to political agendas, may have a significant conflict of interest, especially if they are simultaneously negotiating with the same communities. A key element in these partnerships is a culturally defined society, traditional responsibility; there are both principled and practical reasons to look for autonomous indigenous representational authorities rather than working through nationally defined institutions. Autonomy is a central concern of indigenous societies around the world. The representational authorities claimed by local and national governments for indigenous societies in much, in African societies, still lack credibility. While credibility is, by definition, in the eye of the beholder, credible representative governance systems tend to have participatory processes that can be observed, have representatives that are accountable to community members, and persist over time. The semi-autonomous political status of the most African communities provided them the flexibility under country law to enter into partnerships with international groups of their choosing, despite poorly defined legal rights to the health and research monitoring. Organizations with networks of geographically contiguous communities represented in hierarchical governance systems are likely to be in a better position to represent the interests of the large number of people who can collectively participate in studies than are more dispersed groups.

Although there is some fluidity of community membership between dispersed communities. The existence of a coherent system for assessing and documenting the interests and concerns of communities, resolving differences internally, and authorizing a leader to negotiate on their behalf reduces the transaction costs of developing partnerships and provides for greater project stability (Crigger et al. 2001). RECs partnership is more likely to be able to negotiate with outside partners from a position of strength and withstand challenges from within or without. In fact, many indigenous societies have traditional forms of deliberative democracy for internal decision making, at least at the village level. However, some African communities lack the governance structures to enable this debate to take place at a politically and culturally relevant scale *among and between* communities and to produce credible representational authorities who could translate these discussions into communications and negotiations with Western-style governmental and nongovernmental organizations. The enormous diversity of indigenous cultures and their diverse relationships to Western society and national governments would suggest that the discussion presented above should be treated as a hypothesis and will probably prove to be wrong in many, perhaps most, situations. However, it is fair to say that the involvement of a preexisting and broadly representative indigenous organization will often be important to achieving sustainable collaborative projects of mutual benefit (Foster et al. 1998).

How, then, can global research and development communities respond appropriately to the enormous needs and the scientific opportunities presented by indigenous communities that are neither credibly represented by governments nor organized into larger governance units? Most of the world's indigenous communities do not have autonomous representational authorities. Many of these also have poor representation in their national and municipal governments. While not a completely satisfying response, perhaps the best answer today is that we have a responsibility to help support their capacity to choose development paths that enhance their goals. Researchers have to facilitate the interaction of research and development professionals who work in biomedical and even biological in indigenous communities. However, to the extent that the biomedical health theme involved are communal and the potential for exploitation, real or perceived, is significant, researchers and their funders are functioning in an ethical and political minefield (Jiwa 1999; Macklin 1999). Navigating it will require extraordinary efforts to develop prior informed consent that is based on broad participation and can rely on visible, credible indigenous representation in some form. It will also require that the benefits available to local participants are broad-based and address important local needs that are highlighted by the communities themselves. In some cases preserving local health systems, strengthening environmentally-based health care systems, forming partnerships with local universities, and defining a potential monetary benefit may meet these criteria. In other cases they may not. As academic and industrial research have become more entangled in the past 20 years, biologists, anthropologists and physicians have frequently seized the opportunity to do research that broadened their work from a largely scholarly pursuit to one that might have a more direct impact on applied concerns such as human health through development of new drugs and *in situ* conservation of environmental health through social and economic valuation (Creed-Kanashiro et al. 2005).

To address this issue, professional societies, government sponsors, and academic institutions, hospitals, including non governmental funders, have begun to formalize and publicize policies that clearly separate academic and commercial research. One means of doing so would be to declare that projects that anticipate the possibility of developing a commercial product using genetic or intellectual resources from indigenous societies would be limited to working with societies that have formal systems of accountability and authorization and in which the resources and biomedical attainment can reasonably be ascribed to their authority. The complement to this approach is for purely academic projects to declare their research objectives as strictly for basic knowledge and publication. For the reasons outlined near the beginning of this chapter, this latter approach does not eliminate the potential for exploitation, but it significantly narrows the scope of concern. Biomedical researchers and their institutions should evaluate carefully whether they can do this effectively and ethically in today's politicized climate and their conclusions may vary a great deal between projects. Critical to this approach is thoughtful discussion between researchers and communities about the implications of publication and data basing and what to do if intentions change in the course of a project. Investigators and their sponsors need to consider the possibility of such unintended consequences carefully in project planning and evaluation, but it is worthwhile to place the dynamic in a larger context. While there are many unknowns and widely varying views on the benefits and costs to indigenous communities that collaborate with external organizations, one element of this dynamic is clear. The probability of influence on local community organization is strongly related to the size of the partnership and the nature of the biomaterials and informational resources entering and leaving the community.

Over the past 30 years, research and regulatory agencies around the world have evolved a system of research oversight mechanisms that is increasingly rigid, detailed, and legalistic, but this system has often proved inadequate to protect patients. It has promoted a mentality and practice that encourages researchers to adopt minimum standards of compliance rather than foster a sense of responsibility and thoughtful evaluation by the individuals and groups involved (Morens 1999; Pinching 2000; Zion et al. 2000). Deciding whether a patient understands the potential risks and benefits in an experimental medical treatment and is therefore able to provide *informed* consent is a complex judgment that depends on many situation-specific parameters. Standardizing the information received and the manner in which it is transmitted clarifies the rules of engagement but does not necessarily enhance understanding or ameliorate the underlying inequities in researcher-subject relationships. Some biomedical researchers have very cogently argued that no matter how clearly we establish the rules of ethical research, international projects that involve human subjects in developing countries must be viewed in the context of development. In this context, it is critical to build a culture of responsibility that emphasizes dialogue and collaboration in the global research community and a culture of ethical research in developing countries, including the social, technical, and institutional capacity for appropriate local evaluation and monitoring. However, some argue that the development needs of a host community are frequently a separate question from that which is driving a research project, and it is therefore important that we not create unrealistic requirements regarding the role of the community in project design or the ability of a researcher to address its specific needs. Access and benefit sharing are at least as complicated technically, politically, and ethically as biomedical research with human subjects. Adding the linguistic, cultural, and political layers characteristic of work with indigenous peoples and communal rights produces an extraordinarily complex package that requires tailored approaches to most situations and thoughtful consideration by all involved. There is a need for generally accepted principles, such as those embodied in the CIOMS or the Helsinki human declaration, and national interpretations of those principles in policy measures (Zion et al. 2000). Clear and effective policies are important to give resource "providers" sufficient confidence to seek international collaborations and to give "users" clarity on how to approach the issues and confidence that they are operating within recognized principles and law. The governance of potentially collaborating indigenous societies is key. The generality of this conclusion remains to be seen. Ultimately, conducting ethical research involving indigenous societies will depend a great deal on our ability to help create an enabling environment that includes appropriate local governance systems, well-defined national and international policies, and a culture of responsible research that incorporates locally defined priorities. In such environments we are most likely to achieve what most of us seek- research collaborations that advance science, strengthen human rights, and improve the health and well-being of indigenous peoples and the global community (Deschenes et al. 2001).

Community assemblies for collective decision making

Community assemblies could be employed for collective decision making about initial participation in the project, but formal negotiation of the complex agreements (including benefits) take place outside of the community contexts, among representatives of project partners and their legal counsels. Prior informed consent in legal terms may be about the negotiation of benefits by political authorities, but prior informed consent for the use of environmental, biomedical materials and knowledge within a research ethics framework is not. Further examination of the relational implications of prior informed consent as an ethical principle of "human subject research" may be useful. Prior informed consent is based on the concept of an *ongoing process* that begins with initial contact and continues throughout involvement in the project; it is through this continuing process of communication and information sharing that "subjects" is able to *continue* to choose to participate in research involving themselves. Prior informed consent in the context of research ethics, therefore, is premised upon the existence of a *relationship* between "researcher" and "researched" and is often situated within unequal power relations (Barker and Polcrack 2001). Where research will be conducted with indigenous communities, successful applicants must establish prior informed consent agreements with a preexisting and broadly representative indigenous organization of cooperating indigenous communities (Sobel 2000). When multiple organizations are involved, they must show credible leadership under an umbrella federation capable of arriving at demonstrable consensus on matters of shared concern among participating communities. While perhaps a wise administrative consideration for a government granting agency, this recommendation is prejudicial to the thousands of indigenous communities of the world that are not now and unlikely soon to become members of representative organizations with Western-style systems of governance. It effectively excludes the communities most isolated from national and international markets and most in need of the benefits that would result from their involvement in ethical biomedical programs. Benefit-sharing avoid the pitfalls of bilateral contracts and (a) agreements and implementation are most likely to succeed when the number of parties is minimized, (b) the determination of a competent authority or local focal point in granting access is critical, and ambiguity in this is problematic, and (c) governments need to build local capacity to facilitate the effective and efficient implementation of access and benefit-sharing laws and policies.

The armed conflicts and civil wars affecting indigenous African peoples in much of certain parts of African oblige researchers to broaden the scope to include the national-level politics that ultimately determine the possibilities for development in indigenous communities. With this in mind, it makes even less sense simply to avoid proposing projects in regions where indigenous communities and leadership are inchoate since the environment is really conduce for biomedical research due to neglected health delivery systems. The logic of prior informed consent is that preconceived projects are brought to communities and individuals who then agree to participate. To make this possible, many biomedical researchers use modified, verbal versions of prior informed consent, arguing that written documents simply will not work in politically volatile situations and among semiliterate people (Salinas Mulder et al. 2000; Steinberg 2001). Moreover, individual agreements are not part of the cultural logic in collectively governed communities. But regardless of the legal and liability protections for which prior informed consent is used, until the scientific community reevaluates the philosophy driving the process there may be failures to receive the informed permission that is needed to enter individual and indigenous territories and carry out our jointly constructed and mutually beneficial biomedical projects.

The order of the purpose of collaboration: first, to improve the health and wellbeing of indigenous peoples and the global community and, in the end, to advance science. Seeking prior informed consent of communities can pose challenges so overwhelming that some research initiatives are likely to founder even when key stakeholders agree that their benefits exceed their identifiable risks. Informed consent is quite straightforward as a philosophical concept. In the clinical realm, informed consent should be voluntary, include a discussion of risks and benefits, outline treatment alternatives, and summarize the likely consequences of non-treatment. Informed consent to treatment can be distinguished from informed consent to obtaining biological research samples, but, recognizing these differences, the basic principles of informed consent are the same. Informed consent is a democratizing concept; it is intended to resist the paternalism of professionals and promote autonomous decision making. Social scientists describe many barriers to obtaining genuine informed consent. Some patients express little interest in making health-related decisions. Some research subjects suffer from the therapeutic misconception; they believe that they are receiving beneficial treatment when they are participating in research without guarantee of benefit. Other observers note the complexity of information sheets and informed consent forms, linguistic barriers to communication although currently there is need to translate into local language, differences between the explanatory models of patients or research subjects and health care professionals, the failure of clinicians or researchers to disclose financial conflicts of interest, and differentials in power within the clinical realm.

Other questions about the reliability of informed consent emerge when patients or research subjects express ambivalence about their choices, decision-making capacity fluctuates in response to the severity of illness or the influence of medications, or family members or health care professionals unduly influence deliberative processes and decision making. In short, there is a gap between informed consent as a relatively straight forward philosophical and legal concept and informed consent as a messy, complicated, multifaceted social process. The difficulties associated with seeking prior informed consent of entire indigenous communities, identifies numerous barriers to obtaining prior informed consent of collectives. Abuses from the past can undermine efforts to negotiate agreements involving biomedical scientists, indigenous communities, various academic institutions, and pharmaceutical companies. Histories of colonialism can undermine trust amongst stakeholders. Cultural differences can pose challenges to seeking prior informed consent. Perhaps most important, two major problems will often confront researchers hoping to promote environmental health initiatives. First, the diffusion of knowledge across various groups can raise perplexing questions about who ought to be involved in research initiatives and who can lay claim to being a stakeholder in the research process. The concept of prior informed consent is predicated upon the notion that there is some identifiable collective from which informed consent can be sought. However, defining what counts as a relevant community can be a contentious process. Official bodies established by nation-states do not necessarily have credibility in particular communities. Sometimes various groups are engaged in a struggle for control of local resources or the support of different constituencies. Where the boundaries of groups are well-defined, obtaining prior informed consent can begin with some basic assumptions about whose consent should be sought. In other instances, it is not always clear who ought to be engaged in dialogue.

The involvement of NGOs and other social actors purporting to represent the interests of various groups raises additional questions about whose voices should be given greatest consideration in negotiation processes. Second, even if a community can be identified, it will not always be possible to identify authoritative bodies or leaders who can represent the interests of that community, engage in negotiations, support or question educational initiatives, and sign contractual agreements. Conflicts involving various governing bodies, disputes within these bodies, and questions about their legitimacy can undermine efforts to negotiate agreements with indigenous communities. Governing bodies can disagree about how to weigh risks and benefits and about what constitute appropriate forms of remuneration, capacity building, and local involvement in research activities. As a philosophical concept, informed consent, whether at the level of the individual or of the community, is well defined and easily comprehended. Situated in time, located in particular places, inserted into narratives of colonialism, and connected to the interests, concerns, and local knowledge of particular groups, informed consent as a social process is often remarkably complex, murky, and convoluted. Community informed consent points to asymmetrical power relations and suggest that prior informed consent is an inappropriate framework with which to engage indigenous peoples in this type of research. Each, explicitly or implicitly, points to participatory approaches as more appropriate. However, for better or worse, the terminology of consent and the regulatory framework it reflects are a function of the philosophical and political history of very diverse activities in biomedicine and sustainable environmental development (Jokinen et al. 2002). This regulatory framework currently establishes the most generalizable minimum standards for the conduct of research (CIOMS 2002). The concerns expressed include the difficulty in differentiating basic from commercial research, how to evaluate whether a given indigenous society satisfies this standard, whether this would effectively withhold opportunities from societies that have not developed sufficiently to meet these standards and the possibility that such an approach would preclude projects, such as drug or vaccine research efforts, with great potential for public health. Indeed, it can be difficult to separate basic from commercial research in the early stage of many projects today. Most commercial products in the biomedical world began with a discovery made in a basic research project. Nonetheless, the overwhelming majority of biomedical and environmental research projects involving individuals and communities are driven by basic interests and conducted by scientists with strictly noncommercial objectives (Lavori et al. 2002). Given the suspicion that both greets and follows Western scientists working in indigenous societies today, is it not worthwhile to consider how to minimize that suspicion? Moreover, while we cannot predict the future, we can identify two clear decision points at which basic research becomes commercial. Greater complexities arise when a serendipitous and potentially commercial discovery occurs during a project or subsequent to publication. By considering these issues ahead of time one can be honest about plans and potential outcomes with partners in the developing countries and use this discussion to build trust. But who should evaluate an indigenous organization's authority and by what standards.

Human subjects research protocols are already evaluated to one degree or another, generally by the independent scientific merit review committee, the advisory council of the funding agency, and the agency staff and subsequently by institutional review boards (often at multiple institutions), the local government, and in many cases the community itself. It would be unwise to suggest adding new bodies to this process. Rather it would be wise to suggest that those developing new projects as well as those involved in the above-mentioned review efforts ask themselves whether, given that communities are involved, the collaborating community is competent to authorize the proposed research. The question is analogous to the competence question regularly considered in classical biomedical consent but perhaps more complicated because of the political, cultural, and governance variables relevant to the subjects or resources in question. Because the relevant policies for genetic resources and associated environmental and biomedical research are still evolving in most countries, the answers are sometimes ambiguous. However, asking the question is enormously important and stimulates all the participants and stakeholders to think about the nature of the collaboration, review their approaches, and revise relevant policies and procedures. Would strict application of the competent-authority principle (if standards could be identified) punish less-well-organized societies and prevent important medical or public health discoveries? This is a very difficult question but an extremely important one. In most research ethics decisions, we need to evaluate the potential costs and benefits of a given project to the immediate participants and the global community, as well as the quality of the research, the fairness of the project design, and the competence of the participants to give informed consent.

References

[1] International ethical guidelines for biomedical research involving human subjects. *Bull. Med. Ethics:* 17-23, 2002

[2] Bandewar, S.: Cultural barriers, 'competence' and informed consent in population-based surveys. *Issues Med. Ethics. 11:* 49-51, 2003

[3] Barker, J. H. and Polcrack, L.: Respect for persons, informed consent and the assessment of infectious disease risks in xenotransplantation. *Med. Health Care Philos. 4:* 53-70, 2001

[4] Corrigan, O.: Empty ethics: the problem with informed consent. *Sociol. Health Illn. 25:* 768-92, 2003

[5] Creed-Kanashiro, H., Ore, B., Scurrah, M., Gil, A., and Penny, M.: Conducting research in developing countries: experiences of the informed consent process from community studies in Peru. *J. Nutr. 135:* 925-8, 2005

[6] Crigger, N. J., Holcomb, L., and Weiss, J.: Fundamentalism, multiculturalism and problems of conducting research with populations in developing nations. *Nurs. Ethics. 8:* 459-68, 2001

[7] Deschenes, M., Cardinal, G., Knoppers, B. M., and Glass, K. C.: Human genetic research, DNA banking and consent: a question of 'form'? *Clin. Genet. 59:* 221-39, 2001

[8] Foster, M. W., Bernsten, D., and Carter, T. H.: A model agreement for genetic research in socially identifiable populations. *Am. J. Hum. Genet. 63:* 696-702, 1998

[9] Helgesson, G. and Johnsson, L.: The right to withdraw consent to research on biobank samples. *Med. Health Care Philos. 8:* 315-21, 2005

[10] Jamison, J. R.: Informed consent: an Australian case study. *J. Manipulative Physiol. Ther. 21:* 348-55, 1998

[11] Jiwa, M.: The recruitment of non-English speaking subjects into human research. *J. Med. Ethics. 25:* 420-1, 1999

[12] Jokinen, P., Lappalainen, M., Merilainen, P., and Pelkonen, M.: Ethical issues in ethnographic nursing research with children and elderly people. *Scand. J. Caring. Sci. 16:* 165-70, 2002

[13] Katz, J.: The Nuremberg Code and the Nuremberg Trial. A reappraisal. *Jama. 276:* 1662-6, 1996

[14] Kishore, R. R.: Biomedical research and mining of the poor: the need for their exclusion. *Sci. Eng. Ethics. 12:* 175-83, 2006

[15] Kopelman, L. M.: Minimal risk as an international ethical standard in research. *J. Med. Philos. 29:* 351-78, 2004

[16] Lavori, P. W., Krause-Steinrauf, H., Brophy, M., Buxbaum, J., Cockroft, J., Cox, D. R., Fiore, L., Greely, H. T., Greenberg, H., Holmes, E. W., Nelson, L. M., and Sugarman, J.: Principles, organization, and operation of a DNA bank for clinical trials: a

Department of Veterans Affairs cooperative study. *Control Clin. Trials. 23:* 222-39, 2002

[17] Macklin, R.: Understanding informed consent. *Acta Oncol. 38:* 83-7, 1999

[18] Macpherson, C. C.: To strengthen consensus, consult the stakeholders. *Bioethics. 18:* 283-92, 2004

[19] Marshall, E.: Research ethics. NIH examines standards for consent. *Science. 280:* 1688, 1998

[20] Marshall, P. A.: Human rights, cultural pluralism, and international health research. *Theor. Med. Bioeth. 26:* 529-57, 2005

[21] Molyneux, C. S., Peshu, N., and Marsh, K.: Trust and informed consent: insights from community members on the Kenyan coast. *Soc. Sci. Med. 61:* 1463-73, 2005

[22] Morens, D. M.: Should the Declaration of Helsinki be revised? *N. Engl. J. Med. 341:* 1851; author reply 1853, 1999

[23] Patil, R.: Research in a tribal community. *Indian J. Med. Ethics. 1:* 122-3, 2004

[24] Phipps, E., Fleetwood, J., and Piraino, A.: Research ethics and clinical trials: fostering collaboration with research participants. *Account Res. 6:* 215-26, 1999

[25] Phoenix, J. A.: Ethical considerations of research involving minorities, the poorly educated and/or low-income populations. *Neurotoxicol Teratol. 24:* 475-6, 2002

[26] Pickering, N.: Indigenous peoples and medical research. *N. Z. Bioeth. J. 4:* 2, 2003

[27] Pickworth, E. and Brazier, M.: Fees and research ethics committees. *Bull. Med. Ethics No. 151:* 18-20, 1999

[28] Pinching, A. J.: Live attenuated vaccine trials in medically informed volunteers: a special case? *J. Med. Ethics. 26:* 44-6, 2000

[29] Pinkerton, C. R., Ablett, S., Boos, J., and Philip, T.: Ethical approval for multicentre clinical trials in children. Contrasting systems in three European countries. *Eur. J. Cancer 38:* 1051-8, 2002

[30] Pope John, P., II: The profit motive behind too much medical research. *Origins. 31:* 754-5, 2002

[31] Pope John, P., II: Biomedical research: benefits, limits of science. *Origins. 32:* 673, 675-6, 2003

[32] Resnik, D. B., Zeldin, D. C., and Sharp, R. R.: Research on environmental health interventions: ethical problems and solutions. *Account Res. 12:* 69-101, 2005

[33] Rhodes, R.: Rethinking research ethics. *Am J Bioeth 5:* 7-28, 2005

[34] Saha, S.: Challenges in research in tribal communities. *Indian J. Med. Ethics. 1:* 125, 2004

[35] Salinas Mulder, S., Rance, S., Serrate Suarez, M., and Castro Condori, M.: Unethical ethics? Reflections on intercultural research practices. *Reprod. Health Matters. 8:* 104-12, 2000

[36] Sleeboom, M.: The Harvard case of Xu Xiping: exploitation of the people, scientific advance, or genetic theft? *New Genet. Soc. 24:* 57-78, 2005

[37] Sobel, M. E.: New challenges for biomedical research utilizing human biological materials. *J. Mol. Diagn .2:* 1-4, 2000

[38] Srinivasan, S.: Research on public health interventions in poor countries. *Issues Med. Ethics. 9:* 118-9, 2001

[39] Steinberg, K. K.: Ethical challenges at the beginning of the millennium. *Stat. Med. 20:* 1415-9, 2001

[40] Turner, L.: From the local to the global: bioethics and the concept of culture. *J. Med. Philos. 30:* 305-20, 2005

[41] Wicher, C. P. and Michalek, A. M.: When is informed consent not enough? *J. Cancer Educ. 20:* 9-10, 2005

[42] Zion, D.: Community without communitarianism: HIV/AIDS research, prevention and treatment in Australia and the developing world. *Monash. Bioeth. Rev. 24:* 20-31, 2005

[43] Zion, D., Gillam, L., and Loff, B.: The Declaration of Helsinki, CIOMS and the ethics of research on vulnerable populations. *Nat. Med. 6:* 615-7, 2000

In: A Gateway to Biomedical Research in Africa
Editor: Takafira Mduluza, pp. 61-72
ISBN: 1-60021-444-4

Chapter VI

Protection of Human Subject in Biomedical Research in Egypt

Taha El-khobhy*

Egypt has a history that goes back to almost seven thousand years. This civilization was built on the banks of the Nile River creating a life and a culture, which amazed the world for centuries and still holds a great number of undiscovered mysteries. The Egyptian civilization includes every aspect of life in Science, Medicine, Arts, etc. Because of its unique geographical location between Africa, Asia and Europe, its wealth of natural and human resources, and the potentials it embraces, Egypt has always been a point of attraction to many invaders whose presence entailed the existence of a multi-cultural environment within its boundaries.

Area

Around one million square Km, of these only 55 thousands square Km (i.e.5.5%) is populated.

Population

According to the last census of 2004, population of Egypt reached 71.8 million with a birth growth rate of 1.8 %

* Taha El-khobhy: PhD; VASCERA; Egyptian Organization for Biological and Vaccine Production; Schistosomiasis Vaccine Development Project; 51 Wezarat El-Zeraa St., Agouza, Giza, Egypt; Tel: +202 336 2639; Fax: +202 760 1746; e-mail: telkhoby@yahoo.com

Political Environment

The Constitution of Egypt adopted in 1971 and amended 1980, declares that Egypt is an Arab Republic with a democratic system. The constitution further outlines Egypt's political system and defines public authorities. Executive authority is vested in the Head of State: the President may be elected for other subsequent terms of 6 years. The President formulates and supervises implementation of general state policy. He also acts as a Supreme Command of The Armed Forces. The Government is the supreme executive and administrative organ and council of Ministers headed by the Prime Minister, who supervises the work of the Government. The People's Assembly is the Legislative Branch of the State. It approves the general policy, new laws, the budget and the development plans.

The political system is based on a multi-party system and the formation of political parties is regulated by law. Administratively, Egypt is divided into 26 Governorates; each is headed by a Governor who is appointed by the President. Within their districts, local government units establish and manage all public utilities, provide services and designate industrial areas. Local popular councils are elected bodies that work closely with local government administrative units at various levels.

1. Country Biomedical Research Status

Pharaonic Medicine: Many Papyri describe how medicine was developed to an amazing degree during the pharaonic era. Biomedical research was going hand in hand with medical practice. These researches include general medicine, endemic diseases, ophthalmology, surgery in addition to pharmacology and pharmaceutical industries, botany and herbal medicine. Nowadays the biomedical research in Egypt is a wide and active process. There are about 18 Universities located at various locations in the country. In addition to that, there are a good number of research institutes situated in the big cities, e.g. Cairo, The Capital, and Alexandria and some other cities. There are also some 38 pharmaceutical companies working in Egypt, the majority of them are Egyptians, whether governmental or private companies, the rest are multinationals. All the above-mentioned organizations undergo biomedical research.

Areas of research are wide, variable and responding to different demands through various categories. These areas of research could be classified into:

1. Group of research for the control of infectious and endemic diseases is further classified into chemotherapy and vector control researches.
2. Vaccine development researches started a long time ago for combating infectious diseases in the country.
3. Diagnosis improvement researches, mainly for development of better tools for diagnosis of communicable and endemic diseases.
4. Socio-economic researches to face major public health problems, e.g. Schistosomiasis and viral hepatitis.

5. Epidemiological researches to study the relationship between the parasite (infectious agent), the host and environment and accordingly introduce better control interventions.
6. Health system reform research: this major project is carried by the ministry of health and population. The main goal of the programme is to develop and upgrade the health system in Egypt through upgrading the health system structure and strengthen the referral system between the primary, secondary and tertiary care facilities.
7. Operational researches: this category of researches aims for the development of health activities and overcome any problems arising through application of health interventions.

2. The Study Population and Communities Available for Biomedical Research

The study population and communities available for biomedical research or studies vary greatly in Egypt. As regards beliefs and fears there are wide differences between rural and urban areas. More than 50% of the population live in the rural areas, with their habits and traditions differing from the population of the urban and semi-urban areas, where the socio-economic and political status is much better than that of the rural population. That of the semi-urban might come in between, although the majority of the semi-urban lack the necessary requirements of infrastructures and services.

Habits and traditions in lower Egypt (Nile Delta) vary than those of Upper Egypt. Rural communities in lower Egypt are more developed than those of Upper Egypt, having better infrastructure and higher socio-economic standards. This is due to the availability of wide areas of cultivated land and major projects of land reclamation. The approach of rural populations in Lower Egypt for applying biomedical research is much easier than that of the rural populations in Upper Egypt. Rural populations in Upper Egypt are more conservative than those of lower Egypt due to habits, cultures, state of education, mode of living, illiteracy and socio- economic status. The approach of urban communities is much easier as a whole than that of the rural communities. This is due to multiple factors including better socio-economic condition, literacy, communication - information and higher standard of living. There are key personnel who should be taken into consideration when approaching the communities for doing biomedical research. The first is the medical or health staff of the health centre in the area where the research is going to be applied. The health staff should be involved and informed about the importance of the research, its output, how it may help to overcome a health problem, how it aims to better control or prevent the occurrence of disease. The health staff cooperates with research team in a way or the other for getting the trust of the research community. The second key point is to involve the natural community leaders of the village or district. Those people are popular, having the ability to collect people and convince them about the feasibility and importance of the research.

Clinical Trials and Hospitalization of Volunteers

Researches dealing with the use of new drugs or new devices for management should fulfill the requirements to be used on humans. Clinical trials should be done under strict medical supervision. Volunteers under the trial should be hospitalized in a fully equipped hospital or medical center having qualified and competent medical stuff. The vital functions of the volunteers, e.g. liver function tests, kidney function tests, blood pressure…..etc should be investigated before, during and after the trial according to a special regimen in order to ensure the safety of the drug. There should be all the necessary arrangements to deal with any side effects or complications arising during the trial. These precautions must be written in the protocol of the research.

It should be mentioned that any change in the protocol of the research done by the investigator should be approved first by the IRB, and any deviation from the originally IRB approved protocol without the IRB approval leads to hold and discontinuation of the research.

The Community Hierarchy for Biomedical Research

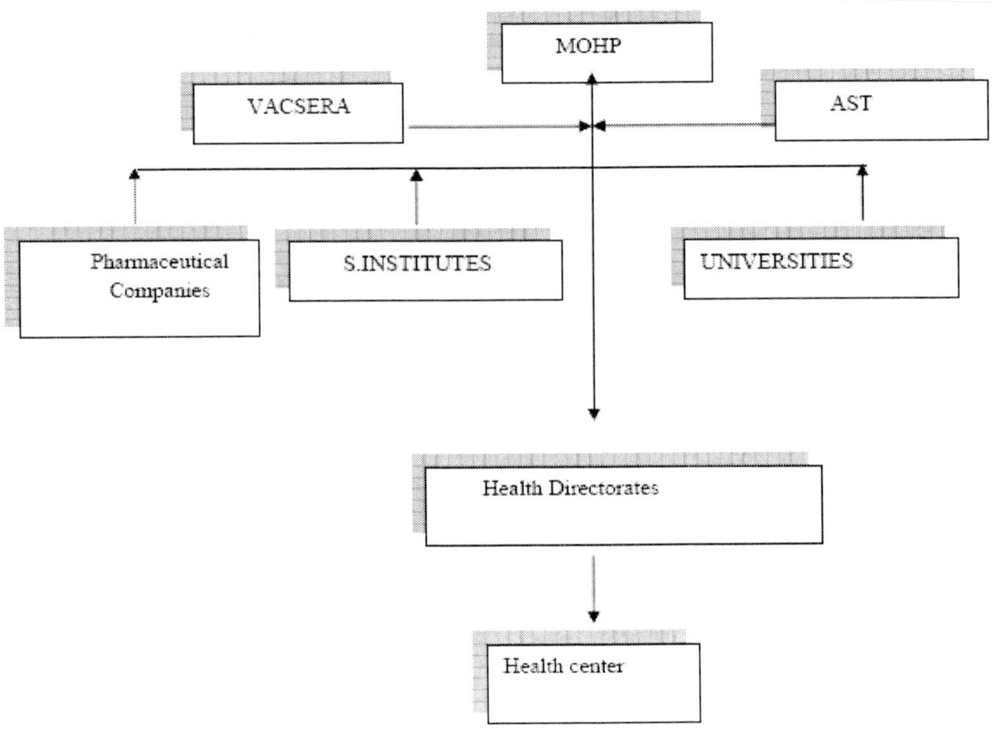

Ministry of Health and Population (MOHP) IRB

MOHP, IRB is the national top review board for evaluation of biomedical research in Egypt. Our constitution states that The Minister of Health and Population is the responsible person for the protection of health and welfare of all Egyptians. Any investigator, researcher to conduct biomedical research or study on humans should take the permission of MOHP. The Permission of the national IRB of MOHP should be presented to the local health authorities in the governorate where the research or the study is going to be done. Accordingly, the governorate health authorities give the permission to the researcher in order to contact the local health center/s to get approval and help to approach the research population.

In addition to the national review board of the MOHP, there are some semi-national review boards located in some major organizations in Egypt. Examples of semi-national review boards are that of the Academy of Science and Technology (AST), and that of the Egyptian Organization for Biological Products and Vaccines (VACSERA). The IRBs of these organizations evaluate the research protocols and studies to be done either by the researchers belonging or not to these organizations. There is a good coordination between the national top IRB of MOHP and the semi- national IRBs. 1n addition, the national top review board located in the Ministry of Health and Population in Cairo, acts as reference for the IRBs all over Egypt.

National Bioethics Processes in Evaluating Biomedical Research Into Country

Almost every university and research institution has its IRB, which evaluates and monitors the biomedical research or study to be done. The independent ethics committee of the institutional review board is constituted and operated according to the rules and regulations as detailed in the Helsinki Declaration and Belmont Report and coincides with the international rules and regulations, as regards the protection of Human Rights in Biomedical Research. Certain criteria should be fulfilled in the protocol of the research or study before submission to IRB:

- Aims/Objectives of the Research:
    1. To study the efficacy aspects of the drug or device to be used in the research.
    2. To report and assess any adverse events following the use of that drug/device.
- Public health importance of the research/study.
- Study/Research location.
- Study/Research time.
- Study/Research forms.

Standardized data collection forms (e.g. case registration form including clinical picture of the subject, dose form, discharge form, follow-up form) should be used to record the data about the efficacy and the adverse events that may be occurred.

Research Ethics Policy

Standards

The IRB employs standards as defined by Declaration of Helsinki, Belmont Report and other relevant international guidelines and rules.

Scientific Merit

The IR B does not primarily rule on the scientific merit of the research proposals presented to it for review. However, the IRB adopts the principle that bad science is unethical. Therefore, there must be scientific value to the research proposal for the IRB to approve it, and the committee will conduct/obtain a scientific review to satisfy itself on this point. Protocols must follow generally accepted standards and Good Clinical Practices guidelines and regulations.

Equipoise

For a research proposal to be ethically acceptable a state of equipoise must exist, i.e. there is genuine uncertainty about the outcome of the research.

Acceptable Research Subjects/Vulnerable Subjects

In general, the IRB accepts studies in patients who are mentally competent and over the age of majority. The IRB also recognizes that drugs or devices often have been approved for marketing by regulatory agencies utilizing data derived mainly from research conducted on relatively young male subjects. Because of scanty data in women, children or the elderly, this situation can pose problems for clinicians who wish to use such compounds in these populations. Therefore, for this reason and in keeping with the Distributive Justice principles, studies that include women and children and the elderly, as well as other classes of vulnerable subjects, are considered to be desirable and necessary as long as subjects are capable of giving informed consent, there is a favorable potential risk benefit ratio to subjects, adequate safeguards are in place and the research is worthwhile.

Informed Consent

The IRB notes that an appropriate informed consent is a central component of human subject protection. The consent process must reflect the first Belmont principle, i.e. Respect for persons, the first application of which is full, free and voluntary consent of subjects. All the regulatory elements for informed consent should be present and acceptable to the IRB.

Version and Patient Initials

Consent forms must be identified on each page, e.g. study number and date and patient initials should be included.

Patient Information and Consent Form (See Annex)

Every consent form should have an introduction including an invitation to participate, brief statement about the study. Study purpose/objectives should be clearly explained. A brief explanation should also be provided of what the study drugs are and how they work. Subject should be told how many research subjects have been exposed to or treated with the experimental device/substance in research studies to date. Every informed consent must have a non-waiver of rights statement, e.g. by signing this consent form you are not giving up any of your legal rights.

Compensation

In addition to medico-legal compensation issues, patients must be informed whether or not reimbursement is available for expenses, attendance at clinic visits, no-charge medication, and etc. Compensation to subjects must not be excessive so as to be an inducement to give consent, and not wholly contingent on study completion.

Risk

Risk and discomforts, including those of study drug, procedures, tests, and placebo doses must be explained in layman's terms.

Risks to confidentiality: Investigators and sponsors have a legal obligation to safeguard the personal health information of research subjects.

Acceptable Standards for Inclusion of Children, Women and the Elderly are:

Children

Clinical research in children is to comply with, and be reviewed under the accepted national and international relevant regulations and rules. Children are defined by IRBs as persons who have not attained the legal age of majority in the jurisdiction in which the research takes place. In most cases the age of majority is 18 years. Minors judged sufficiently mature to read or understand simplified version of the consent form.

Assent

Generally accepted age of assent is nine years of age and over. Research involving younger children and infants may also be acceptable, so long as the research holds promise of possible therapeutic benefits, important information may be learned as a result of the research, and either risk is minimal or sufficient safeguard are in place to afford appropriate protection to children.

Women

Women of any age who are not nursing should be included in the research, unless specific justification is provided for their exclusion. Women who are capable of childbearing and sexually active must practice an acceptable method of birth control that will not interfere with the research or interact with the research drug.

Elderly

Patients over the age of 65 years who are capable of giving their own consent and are able to comply with the requirements of the study, provided that data is available regarding possible need for dosage modification, drug interaction, to show that the risk/benefit ratio of the proposed research in the elderly is acceptable.

Vulnerable Populations

Vulnerable populations of research subjects include but are not necessarily limited to the following groups:

- Women, especially if pregnant or nursing, women of childbearing capacity.

- Children (persons under the age of majority i.e.18) especially neonates, infants and young children.
- Geriatric population.
- Adults with diminished mental capacity.
- Economically or socially disadvantaged persons.
- Persons with certain physical disability.
- Illiterate persons.
- Prisoners.

Justice Principle

The Justice principle states that individuals from those groups that may benefit from the results of research have an obligation to assume the risk of participation. Therefore, justification should be provided for the exclusion of any group that in the past, was arbitrarily excluded from research participation simply because of vulnerability of that population.

Health Care System Utilization Costs

The IRB requires some form of assurance from either the Sponsor or the Investigator that health care system utilization charges as a result of study requirement, e.g. subject visits and laboratory tests, be borne by the sponsor and that the investigator will not bill the public system for costs covered by the study budget. A statement may appear in the investigator agreement whereby the investigator agrees in writing not to bill the public health system for services that are to be covered by the grant.

Monitoring of Biomedical Research

Ethical commitment and protection of human rights is the most important aspect in Biomedical Research Monitoring. All the ethical commitments regarding human rights should be fulfilled. Any deviation from the original proposal should be described, clarified and justified by the Principle Investigator and approved in written by the IRB in advance. The PI should submit periodic reports to the IRB which meets periodically to review these reports in order to see whether the research process is following the laid down ethical procedures as mentioned in the original proposal as well as according to the IRB set of ethical standards.

Progress Report

Where an IRB approval is about to lapse and an extension to the approval is necessary to complete the study, an application for approval extension is required. The application must be supported by a report that contains information as described in the Investigator Progress Report Form issued by the IRB to investigators upon their approval.

End of Study

When a site is closed, an end of study report must be filed with the IRB on a timely basis. Whenever possible, the IRB would appreciate a summary of the entire trial from the sponsor.

Dissemination of Research Result of Clinical Trials

IRBs strongly encourage that there will be timely dissemination of results, may be by means of clinical trial registration, presentation at scientific meetings, journal publications, or other appropriate means, preferably within six months but not more than one year of final data analysis.

Current and Future Status of Biomedical Research in the Country

Rural/Urban community based study or research is a process involving research team, research community (participants or volunteers), medical staff of the health center in the area of research and the community leaders in that site. Usually there is good cooperation between all these groups, due to mutual understanding about the importance of the research, and the community could benefit from the results and outcome of the research. The medical staff of the health center involved in the research usually would be compensated financially according to the time they spend on the research. Almost all researches have in the budget a component for that. The research community usually gets some in kind incentives and transportation fees if there is any. In clinical trials needing hospitalization, the participants may get financial compensation in addition to free medical care. Egypt is rapidly growing into a medium income country. It is one of the political stable countries. There is a strategic development plan, 5 years development plans working since more than 20 years ago. The Egyptian community is considered to be a stable community. There is no emigratory activities, no ethnic troubles and the family bond is strong.

The political and community stability enhances and supports biomedical research activity in Egypt. In this regard we can not neglect the cultural aspects of the Egyptians which have positive contribution on the biomedical research activities in the country. It should be noted that research team is composed usually of Egyptian researchers. This fact enhances the encouragement and co-operation of the research population.

The single language of the researchers and the research community facilitate mutual understanding between either side. In addition, the research community better trust the researchers without any fears or suspicions. In some cases, the research team might have a foreigner scientist in which case the majority of the team and the principal investigator are

Egyptians, and the language used is that of the research community i.e. Arabic, which alleviates any worries or fears that might arise. In Egypt, there are some areas of research that need more consideration in the future. Of those areas; zoonosis is one. Zoonotic diseases are numerous in the Egyptian community especially in the rural areas. Another example is that of imported diseases, whether infectious or parasitic diseases especially those diseases or vectors of diseases invading the southern borders of the country.

Conclusion and Future Entry Requirements Into Community

Biomedical research is an active process in Egypt, dating back to the pharaonic times. Almost all universities and scientific institutions have ethical committees of the IRBS. The top national IRB is within the Ministry of Health and Population. In addition there are two semi-national IRBs, one at the Academy of Science and Technology and the other at VACSERA. The ethics review committee of the IRB limits itself to research on human subjects. The committee is absolutely independant, a stand-alone entity not founded or related to any governmental or non- governmental organization. The committee operates in accordance with Belmont Report, Declaration of Helsinki and other national and international laws and standards.

Protocols Are Reviewed Periodically by the IRBS

The ethical committee of IRB encourages the investigator to inform and involve the Family Physician or the Senior Health Center Physician in the subject's selection and participation in any clinical research or study. This is because it is possible that the family physician and the health center physician are aware of factors that may be important in the selection process and suitability of the participant to participate in the study that are unknown to the investigator. Furthermore, it is likely the subject will return to him for continuing care. The inclusion of the regular physician in the study should be based upon permission of the participant subject.

References

Abdel Wahab F. M., Yosery A., Narooz S., Esmat G., El Hak S., Nasif S., Strickland G.T. (1993) Is Schistosoma mansoni replacing Schistosoma haematobium in the Fayoum? *Am. J. Trop. Med. Hyg.* 49: 697 – 700

Abdel- Wahab MF, Zakaria S, Kamel M, Abdel- Khaliq MK, Mabrouk MA, Salama H, et al. High seroprevalence of hepatitis C infection among risk groups in Egypt. *Am. J. Trop. Med. and Hyg.*1994, 51:563-567.

Abu El Yazeed R.R. Mansour N.S., Youssef F.G., Boghdadi A.M., El Khoby T.A., Hassanein Y.A., El Gamal R.R. (1998). Seasonality as a determinant of the efficacy of Praziquantel in population-based chemotherapy: lessons from the practice. *J. Egypt. Soc. Parasitol.* 28: 1-7.

Abu El Yazeed R.R., Youssef F.G., Merrell B.R., El Gamal R.L., El Khoby T,A, Hassanein YA., Cummings C.E (1996). Paraziquantal in the treatment of Schistosoma mansoni infection: comparison of 40 and 60 mg/kg bodyweight regimens. *Am. J. Trop. Med. Hyg.* 56:404407.

Arthur RR, Hassan NF, Abdallah MY, el- Sharkawy MS, Saad MD, Hack- bart BG, et al. Hepatitis Cantibody prevalence in blood donors in different governorates in Egypt. *Trans. R. Soc. Trop. Med. Hyg.* 1997, 91: 271-274.

Barakat R., El Masry A.G Faraghaly A., El Morshidy H.N., El Sayed M.K., Husein M.H., Miller F.D. (1995). Impact of population-based selective chemotherapy on prevalence and intensity of Schistosoma mansoni infections in the Nile Delta: Kafr El Sheikh. *Trop. Med. Hyg.* 41, 56-62.

Darwish Ma, Faris R, Darwish N, Shouman A, Gadallah M, El-Sarkawy MS, et al. Hepatitis C and cirrhotic liver disease in the Nile delta of Egypt : a community based study. *Am. J. Trop. Med. Hyg.* 2001 ; 64:147-153.

El Khoby T., Watt S., El Katcha S., Schistosomiasis in Egypt: An Update. *The Nigerian Journal of Parasitology,* 12 (1991) pp 51-57.

El Khoby T., Galal N. and Fenwick A.(1998. the USAID/Government of Egypt's Schistosomiasis Research Project (SRP). *Parasitol. Today,* 14. 92-96.

El Sayed H.F., Rizkalla N.H., Mehanna S., Abaza S.M., Winch P.J. (1995). Prevalence and epidemiology of S. mansoni and S. haematobium infection in two areas of Egypt recently reclaimed from desert. *Am. J. Trop. Med. Hyg.* 52: 194-198.

Nafeh MA , Medhat , A, Shehata M, Mikhail NN, Swifee Y , Abed hamid M, et al. Hepatitis C in a community in upper Egypt : I. Cross sectional survey. *Am. J. Trop. Med. Hyg.* 2000; 63: 236-241.

Talaat M. and Evans D (1996). Costs, benefits and operational implications of using quantitative techniques to screen for Schistosomiasis haematobium in Egypt. Southeast *Asian J. Trop. Med. Public Health* 27, 29-35.

Talaat T. M. , Miller FD (1998). A mass chemotherapy trail of Praziquantel on Schistosoma haematobium endemicity in the upper Egypt. *Am. J. Trop. Med. Hyg* .59: 546-550.

In: A Gateway to Biomedical Research in Africa
Editor: Takafira Mduluza, pp. 73-80
ISBN: 1-60021-444-4

Chapter VII

Biomedical Researches in Sudan

*Nagla Gasmelseed**

Introduction

Sudan is the largest country in Africa, it is about 2.5 million sq km^2 extending from latitude 4 to 22 degrees north and from Longitude 22 to 38 degrees East. It is characterized by a strategic geographical location, that links the Arab world to sub-Saharan Africa, and it shares its borders with 9 countries, where the Sudanese population and those of the neighbouring countries move freely across these borders. The environment ranges from damp rainy in the south, to desert in the northern areas. Sudan is classified as a low-income country. It is widely believed that the incidence of poverty is very high and that there is a considerable variation in poverty levels between states. However, Sudan is rich in terms of natural and human resources, but economic and social development have been below expectations.

The Sudan population according to the last census is about 25.6 million while last estimation in 2004 about 34.5 million with growth rate of about 26.3, the birth rate of about 37.8 and the mortality rate 11.5 per thousand population. For the 2004 data shows that the urban population of about 34.5% form the total population within the age (6-24 years old); about 44.5% are less than 15 years old and about 41% are less than 5 years old, while about 15.6% are more than 60 years old from the total population. This information shows that about half of the population are children and older which indicates that there is imbalance in the population structure in addition to the immigration from rural to urban which leads to the increased demand of health and education services.

* Nagla Gasmelseed: Department of Molecular Biology; Institute of Nuclear Medicine, Molecular Biology and Oncology; University of Gezira, P. O. Box 20 Wad Medani- Sudan; Tel: 0024951161593 – 0024951143232; Fax: 0024951153427 – 0024951146640 ; Email: Nag_la@yahoo.com

Administrative Health Structure

The introduction of federalism in Sudan in 1993 fostered a three-layered health system structure. These are Federal, State Ministries of Health and Local health system. The FMOH (Federal Ministry of Health) with its 10 general directorates became the main layer for policymaking, strategic planning, co-ordination, regulation, international relations and central source of technical support and guidance for the states. These roles are reflected in the organizational structure for the FMOH.

Decentralization was introduced in the mid-1990s, where responsibility for government health system management was devolved to the States and localities. The Federal Ministry of Health is linked to 26 State Ministries of Health (16 of which are in the geographic north of the country). Within each State there are a number of localities (134 in total) managed through Health Area System; however less than half of the localities have a functioning Health Area System, and only 19 are reportedly working to the standards. Local councils are also responsible for water and sanitation services. In addition to the Ministry of Health services, there are some university hospitals that are run by universities, as well as other sectors like police and military services. In addition, there are privately run clinics and hospitals.

The 2nd layer, composed of 26 State Ministries of Health, shares the responsibility of planning, legislation and financing with the Federal MOH. However, it takes the direct responsibility for the organization of health in the state and support of the local health system. Nevertheless, due to their weak capacities there are notable gaps. The exceptions are Khartoum and Gezira SMOH, which have relatively better performing health systems.

The third layer in The Federal setup is the locality level. The local health system is based on the district health system approach. It emphasizes the principles of primary health care represented in decentralization, community participation, intersectoral co-ordination and integration of services. The health area system has been established to strengthen the health management capacity in geographic areas around rural hospitals.

Biomedical Research Status

The main importance of the biomedical research is to provide the scientific information which then assist the health worker, politician and health director in the different level to provide good health services which lead to improvement of community health. The priority of research depends on the need of each state and classified in infectious diseases such as malaria tuberculosis, schistosomiasis, meningitis and HIV. Other priorities for non infectious diseases are cancer, diabetes and thyroid in addition to the health informatics system which gives the base line data of different disease. The high priority of biomedical research is malaria, child health and communicable diseases. These research activities have been done in coordination between ministry of health and universities and other academic sectors. There are different sectors that participate in the implementation of the biomedical research; one sector is federal ministries of health, universities and research centers, National Counseling for research WHO and UNICEF UNFPA, the donation depends on the type of research. Other sector is implementation of the research such as National Health director, National Program of Health Faculties of Medicine and health sciences, medical postgraduate studies and national health laboratories.

The high priority of research in the central Sudan is malaria, most of the biomedical research has been built around determining of new technique for control and prevention and to study the behavior, attitude and knowledge in the diagnosis and treatment of the severe form of malaria, evaluation of the different methods in diagnosis of malaria and the prophylactic dose during pregnancy. The medium priority is tuberculosis. The most common study is DOTS and also immunity against TB toward the drug used in Sudan and the reason of the TB patient relapses.

Evaluation of mantoux test in diagnosis of TB in children. Also the study on the efficacy of the vaccination in protection of tuberculosis and the effect of lung endoscopy in diagnosis of tuberculosis. Tuberculosis is a high priority in the eastern Sudan. The biomedical research is supported by the National program for controlling tuberculosis, with major donation from commercial banks, ministry of finance and African organization for research. HIV has a high priority in Southern Sudan in spite of the other infectious disease.

Priority Health Programmes

Programmes for Maternal and Child Health

Reduction of maternal mortality was incorporated as a national priority in all strategic long and short plans since 1992. Making pregnancy safer (MPR) national plan sets a national target, to maximize the proportion of deliveries attended by skilled persons and to avail comprehensive and basic emergency obstetric care in all hospitals. In 1996 the FMOH changed its Maternal and Child Health/Family Planning (MCH/FP) Directorate to the Federal Reproductive Health Directorate (FRHD) under the hierarchy of the Directorate General of Primary Health Care to be responsible for reproductive health services in the country. FRHD is charged with the design of general policy, work plans and follow-up of implementation, laying the principles of collaboration between all sectors involved in reproductive health and monitoring and evaluation of reproductive health activities in the country. On average only just about less than half (43.6%) of all health facilities in northern Sudan offer reproductive health services. While some reproductive health services are seriously lacking (EmOC, standard protocols, adolescent services, community awareness sessions, care of abortion and AIDS services, none of those provided is offered by more than one-third of the available reproductive health facilities.

Child health

Primary health care is one of the main directorates of the Ministry of Health within which most child health programmes, namely EPI, nutrition, IMCI and others are delivered. In addition, there are other programmes, which have indirect impact on child health, for example basic development needs (BDN). Both The Control Departments' of diarrhoeal and the respiratory diseases were merged to form the Integrated Management of Childhood Illnesses Department.

Immunization of children: the FMOH is responsible for strategic planning, supply of vaccines and injection equipment to the states, training of trainers and supervision and monitoring of state level immunization programmes. The state level MOHs provide immunization services through three strategies, i.e. fixed centers at health facilities, outreach services and mobile teams for hard-to-reach areas. Although there are wide inter-state variations, fixed centers on an average are supposed to cover about 50% of target population, outreach teams 20% and mobile teams 30% of the target population. The immunization programme, in terms of vaccines and injection equipment, is fully funded. According to the WHO assessment report of 2000, the major cold chain at state level is in good functioning condition.

Control of Communicable Diseases

Communicable diseases dominate the health scene with high vulnerability to outbreaks.

Malaria: Malaria in the Sudan is a major public health problem. The whole country is considered endemic with varying degrees from hypo-endemic to holo-endemic areas. The country shoulders roughly 50% of all malaria cases in EMRO and with a conservative estimate, there are about 7.5 million cases resulting in 35,000 deaths per year. The disease accounts for about one fifth of outpatient cases and nearly one third of in-patient cases. Malaria is also believed to be the cause of one fifth of all hospital deaths. Malaria case fatality rate for paediatric hospitals ranges between 5 and 15%. The Sudan has a long history of malaria control activities dating back to pre-colonial times. The attempt at malaria eradication started 1954-1964 with the introduction of DDT and it focused on vector control, residual spraying, environmental management and public education. The program had very limited success due to managerial, technical and financial constrains. The establishment of the Blue Nile Health Project (BNHP) in 1978 with support from WHO, WB, Kuwait, Japan and USA, led to successful control of malaria for 10 years with the prevalence of the disease coming down to <1%. Unfortunately, external funding stopped, leading to discontinuation of control operations in 1989.

The WHO announced Rollback Malaria (RBM) in January 1998 as a global initiative to reduce the size of the problem in all countries especially in Africa; through participation and integration of the different levels. This initiative is adopted by the FMOH through the National Malaria Control Programme (NMCP). The overall goal of RBM in Sudan is to reduce the malaria morbidity and mortality by half by 2010 and the general objective of the RBM strategic plan in the Sudan is to halve the malaria morbidity and mortality by 2010.

Schistosomiasis: is one of the main endemic water-borne diseases in Sudan. The disease is wide spread in Gezira and Central states; however, it is endemic in all states of the Sudan except Red Sea. Twenty four million people are under the risk of contracting the infection. The prevalence ranged between 28-80% among school children surveyed in central States in 2001. Since the cessation of the Blue Nile Health Project, there were no organized control activities. The national control programme is still weak with irregular and insufficient funds. However, in line with growing political commitment, partners started to show more interest.

Tuberculosis: An estimated annual risk of 1.8% which gives an incidence of 90/100,000 smear-positive cases, puts Sudan among the high prevalence countries for TB in the East Mediterranean region. The National Tuberculosis Control Programme (NTP) was launched in 1974. Since 1995 NTP is gradually implementing the TB programme at all levels of administration within the health sector in accordance with IUATLD and WHO approved criteria. The NTP implemented the effective 8 month TB treatment SCC-DOTS and managed to achieve *DOTS ALL-OVER* by the end of the year 2002. Case detection rate is 56% and success rate, has reached 75.5%. Reliable data on drug resistance are not available. Reference laboratory conforming with the criteria of the WHO and IUATLD has been installed by NTP/LHL.

HIV/AIDS: The HIV epidemic has grown at an unexpected high and alarming level. High illiteracy rates (especially among females), unhealthy lifestyles and large numbers of IDPs all adversely contribute to the spread of the disease. The cumulative number of reported HIV/AIDS cases in Sudan has reached 11,500 by the end of 2004. Transmission is predominantly through unprotected heterosexual intercourse. Based on recent epidemiological surveys, the national HIV/AIDS control programme estimates put HIV prevalence as 1.6% with 500-600,000 people living with HIV/AIDS.

The Government of Sudan undertook a situation analysis on scientific lines on 2002. The situation analysis helped in finalizing strategic framework for the coming five years and development of an evidenced based implementation plan for five years (2003/08). The UN agencies provided technical assistance in achieving these milestones. Theme group on HIV/AIDS and UNAIDS in Sudan have also embarked on advocacy and sensitizations campaign in the last two years. The FMOH also held an Inter-country meeting for the Ministers of Health of nine neighboring countries to discuss cross-border issues including HIV/AIDS.

Leishmaniasis: is endemic in Sudan, mainly affecting 6 states in the North and South. Visceral leishmanisasis accounts for high mortality in those areas. The yearly disease burden is estimated at 6,000 - 9,000 in the North and 2,000-5,000 in the South. Successive outbreaks occurred during the last few years leading to a lot of mortalities.

Guinea Worm: Sudan remains one of the few countries where the disease has not been eradicated. The program started at the end of 1992, since then, a considerable reduction of cases in the controlled areas in the southern region was achieved. The civil strife in the south which led to continuous displacement towards the north, long duration of the rainy season, difficulties in accessibility and lack of basic infrastructure in the southern states delayed the effort of eradicating the disease from the country.

Sleeping sickness and lymphatic filariasis: are widely prevalent in the Southern States, South Darfur, Southern parts of the Blue Nile State. There is no recent documented information on the prevalence of these diseases or the impact on health in endemic areas. The control programmes for these diseases are weak and under funded.

Blindness Control Programme: Cataract is the leading cause of blindness in the country, it is estimated that 1-1.5% of the population is blind, i.e. 300,000 and are in need of at least one eye surgery which is far from the current capacity: 30-40 thousands cataract extraction operations annually.

Non communicable diseases (NCDs): In late 2001, the Health Education Department in the FMOH was changed to health promotion directorate as an umbrella for seven programmes, namely mental health, school health, oral health, cancer prevention, Sudan initiative for tobacco control, national diabetes program and health of the elderly programmes. However, integration has not yet been fully achieved. Recently, a department for NCDs was established within the PHC directorate.

Health Information System (HIS): A new integrated system based on community information is developed and the process of its implementation has already started.

Research: The role of research in ensuring that evidence-based feedback informed decision-making has not been fully exploited. The Research Directorate was established in 1998 with WHO support to lead research into health systems. The unit initially conducted training programmes in some states and identified priority research issues. The unit is responsible for reviewing and endorsing proposals for research funding.

The effects of the war on health, nutrition and population have undoubtedly been significant. More than 2 million people are claimed to have died, and more than 4 million are internally displaced or have become refugees. A comprehensive peace accord was signed in 2005 ending the civil war in the south. Since February 2003 there has been an escalating military conflict in Darfur. Overall health indicators in Sudan, are comparable to Sub-Saharan Africa averages. However, overall averages mask significant urban-rural and regional disparities, related to conflict, displacement, and chronic poverty. Communicable diseases dominate the health scene with high vulnerability to outbreaks. The trend of none communicable diseases has seen observable increase.

The Community Hierarchy for Biomedical Research

In each research institute there is ethical clearance committee (ECC) which approve the research proposal, after that other ethical committee with responsibility to review the proposals and its priority to the community. Before entry into the communities one needs to get permission from the district or regional administrative MOH. Working with local researchers the team is introduced and presents the ethical clearance certificate, explains the aims of the study and requests a list of all the wards, villages and hamlets available in the district.

National Bioethics Processes

A proposal submitted with the CVs of the investigators to the research administrative unit in FMOH. There are two committees, one for the ethical clearance and the other for the national endorsement. The national endorsement committee is the top of the most review board which includes members from various higher learning institutions in the country, Ministry of Health, including layman. They have a meeting monthly to review the proposals according to priorities. After that the proposals pass through the ethical committee for the ethical approval and endorsement.

References

CBS, UNFPA (2004) Population Data Sheet 2004

CBS, UNFPA (2005) Population Data Sheet 2005

CPA and Machakos protocols on power and wealth sharing.

Decaillet, F., Mullen, P. D., Guen, M., (2003) "Sudan Health Status Report – Draft Version 1," World Bank/AFTH3

GOS (1992) Sudan Comprehensive National Strategy 1993-2002 (health development strategy)

Human resources Department, (2004) " Human resources Strategic plan (2004-2013)" Federal Ministry of Health, Khartoum

Joint Assessment Mission Northern Sudan, Health Sector 2004

Ministry of economic and National Planning, (1990) Sudan Demographic and Health Survey (DHS, 1990)

Ministry of finance and national economy (http://mof-sudan.com/English/sudanese_economy.htm)

Ministry of Finance and national economy (http://www.moe-sd.com/a/s.doc)

Muhammad B.H. (2003) Review of millennium development goals for health, nutrition and population, July 2003, Sudan

National Health Information Centre, (1989) "Annual Health statistical Report 1989," Federal Ministry of Health, Khartoum.

National Health Information Centre, (1995) "Annual Health statistical Report 1995," Federal Ministry of Health, Khartoum.

National Health Information Centre, (1993) "Annual Health statistical Report 1993," Federal Ministry of Health, Khartoum.

National Health Information Centre, (2000) "Annual Health statistical Report 2000," Federal Ministry of Health, Khartoum.

National Health Information Centre, (2003) "Annual Health statistical Report 2002," Federal Ministry of Health, Khartoum.

National Health Information Centre, (2004) "Annual Health statistical Report 2004," Federal Ministry of Health, Khartoum.

National insurance Shem act, 1994 and 2000

Omer, HS., et al., (2001) " PHC in-depth review in Sudan" Federal Ministry of health, Khartoum,

Saad Eldin, El Mukhtar M: Development of Health Service Indicators for Health Areas; 2000-2001

Sudan and UNFPA (2001) National Safe Motherhood Survey: Sudan, 1999, Khartoum.

Sudan and UNICEF (2002) " Multiple Indicator Cluster Survey, 2000 Sudan – Final Report," Federal Ministry of Health and Central Bureau of statistics, Khartoum, available at http://www.childinfo.org/MICS2/newreport/sudan1/sudan1.htm

UNDP (2002) United Nations Development Assistance Framework (CCA/ UNDHAF) (www.sd.undp.org/Doc/CCA-UNDAF-Apr2002.htm)

UNDP (http://hdr.undp.org/statistics/data/indic/indic_4_2_2.html)

Value-Added tax regulation 2000, Taxation Chamber

World Bank. (1990) Making adjustment work for the poor: A framework for policy reform in Africa. Washington, D. C.

World Bank (2003) " Sudan Stabilization and Reconstruction-Country economic Memorandum" available at http://www1.worldbank.org/wbiep/decentralization-africa/CEM%20Volume%20I.pdf

In: A Gateway to Biomedical Research in Africa
Editor: Takafira Mduluza, pp. 81-89

ISBN: 1-60021-444-4

Chapter VIII

The United Republic of Tanzania, East Africa

Stephen E. D. Nsimba[*]

Brief Introduction on Country Biomedical Research Status

Tanzania is a large country with about 950,000 sq km. However, Tanzania is among the ranked poorest countries in the world. It has a population of about 35.6 people according to the national census of 2002 (ratio is 18 Females to 17 males). The majority (75%) of these people live in rural areas. They earn their living through tilling the land using hand hoes by practising what is termed as subsistence agriculture which is in most cases extensive with relatively low outputs. The Tanzanian government have been and continues to struggle hard to fulfil the global call for bridging equality, poverty alleviation, ignorance and diseases. Diseases, Poverty and ignorance are the three major enemies of development in Tanzania, and the government is in the fore front in fighting these 3 enemies in order to build a healthy nation.

Currently, priority research areas in the country are divided mainly into two broad categories depending on the magnitude of the problem; national policy commitments; relevance and urgency. Thus, the two broad main research areas include Biomedical Research and Health Systems Research (HSR). These two broad research areas are further divided into high, medium and lower priority.

1. The Biomedical Research: this includes the following disease topics which are considered as high priority; communicable diseases; major communicable diseases;

[*] Stephen E. D. Nsimba: Muhimbili University College of Health Sciences (MUCHS), Department of Clinical Pharmacology, P. O. Box 65010, Dar-es-Salaam-Tanzania. Tel: 255-22-2150302-6 Ext 214; Fax 255-22-2150465 Mobile: 255-713-318113; Email: snsimba@muchs.ac.tz or nsimbastephen@yahoo.co.uk

neglected diseases and maternal and child health. The medium priority include; disease control; non-communicable diseases; nutritional and basic research; whereas the lower priority include; environmental health; product development; gender; traditional and alternative medicine and occupational health.

2. The Health Systems Research: this include the following disease topics which are considered as high priority; human resources for health; reproductive and child health; health service delivery; HIV/AIDS and health financing. The medium priority include; drugs and medical supplies; health information; health policy; essential health interventions packages and decentralisation; whereas, the lower priority inter-sectoral collaboration; public-private partnership and international funding initiatives.

However, the health determinants and socio-cultural issues are recognised as "cross-cutting" set of issues which is considered under both the biomedical and the health systems research. These Health Research Priorities are used to guide and inform the development of research portfolios and research investment choices for Tanzania for the next five (2010) years.

The Study Population and Communities Available for Biomedical Studies

Tanzanians are generous people (be urban or rural communities) and are always cooperative and ready to participate in any beneficial biomedical research study provided it is well explained in advance through their immediate leaders. However, fears and beliefs in some rural communities may be there especially if the research team may only be comprised of foreigners which actually are not likely to happen. Any foreign researcher (scientist) usually (advised to involve Tanzanian researchers) involves at least a Tanzanian researcher and picks field / research assistants who are Tanzanians and for that matter are familiar with the Swahili language which is the national language used as a medium of communication by all Tanzanians in rural and urban areas in the country. The situation may be scary to rural people especially if the research involves taking human biological samples such as blood (blood slide for malaria parasites, blood samples for assay, urine etc) if not properly explained by their local leaders and the research team may cause a lot of fear and some believe that researchers are taking such samples for screening of their HIV sero-status without their consent.

The economical and social status of Tanzanian people varies from place to place especially in the rural communities and partly this is contributed by the different climatic conditions Tanzania experiences. For example the northern parts of the country (e.g. Arusha and Kilimanjaro), the southern parts (Ruvuma, Iringa, Mbeya and Rukwa), Lake Zone (Mwanza and Kagera) regions these regions have a fertile land and produce a lot of food as well as cash crops. So economically, people in these rural areas are usually relatively far better when compared to other rural communities within the country.

The United Republic of Tanzania got its independence in 1961 and the union between Tanganyika (mainland) and Zanzibar (island) in 1964 under the leaderships of the late

President Mwalimu Julius Kambarage Nyerere and Abeid Aman Karume respectively. Tanzania has been and continues to be one of the peaceful and stable (political status) country in Africa. That is why we are seeing the various peaceful transitions of handing over of presidential power for the fourth time in a peaceful and respectable way since the introduction of the political multi-party system in 1997 to date with more than 16 political parties. However, despite the fact that the country has peace and stable government, the country and its people face some natural disasters such as lack of enough rains in the country, floods in cases when we receive too much rain etc. These calamities of weather contribute to food shortages in the country. In this regard, food supplies do not match food demands. The development of agriculture has been an important objective of the government. The focus has been on producing more food to enhance food security and alleviate poverty so that nutritional status of the people is improved as well as the economic well being, with the ultimate goal of becoming self-sufficient in basic food requirements and also to sell some excess for both local cash generation and foreign exchange earnings.

The Community Hierarchy for Biomedical Research

Upon receiving Ethical Clearance Certificate from the Medical Research Coordinating Committee (MRCC) or Tanzania Commission for Science and Technology (COSTECH) for any research in which the Principal investigator is non –Tanzanian or Tanzanian, they need to ask permission to get into the district or region from the District (DC) or Regional (RC) Commissioner. The respective DC or RC will provide them with a letter to introduce the research team to the Local Government Leaders of the respective study site (s). Here you need to have a copy of your Ethical Clearance Certificate with you which has been issued by the MRCC of the National Institute For Medical Research –Tanzania or from Tanzania Commission for Science and Technology (COSTECH).

However, before you enter into the communities you need to get permission from the district or regional administrative authorities. Here I will present a scenario from my studies in rural communities of Kibaha district, Coast region- Tanzania how we entered into the communities. During logistic preparations prior starting the actual study we visited the District Commissioners office (DC), and we were assigned to see the District Executive Officer/Director (DED/DEO) on behalf of the DC. We introduced ourselves and presented our ethical clearance certificate, explained the aims of our study and requested him/her to get a list of all the wards, villages and hamlets available in his/her district or region.

The DED/DEO also can ask questions in case he/she wants to know more about the study. Then he writes a letter of introduction to all his/her ward executive secretaries/officers. Then you take these letters physically for simplicity of the process to the ward secretaries/officers who again you meet them and talk to them and tell them all your study plans. These Ward Executive Secretaries (WES/WEOs) would again call a meeting of the village and hamlet secretaries or chairmen. If you want you can arrange with the ward secretaries another day to meet with the village and hamlet secretaries and chairmen so that you brief them all your study plan and allow them to ask you questions in order to have a clear picture about the study and how it will be carried out (logistics) so that they can inform their own people in their villages or hamlets. By definition a "hamlets" is the smallest administrative unit in a village or city. Here a good study plan is needed so that you don't interfere so much with peoples daily activities (i.e. when you promise them you will visit every say Tuesday and Wednesdays, you must fulfill your promise) as people will leave all their personal developmental activities to wait you researchers.

Generalized Summary Diagram on the Hierarchy for Obtaining Permission

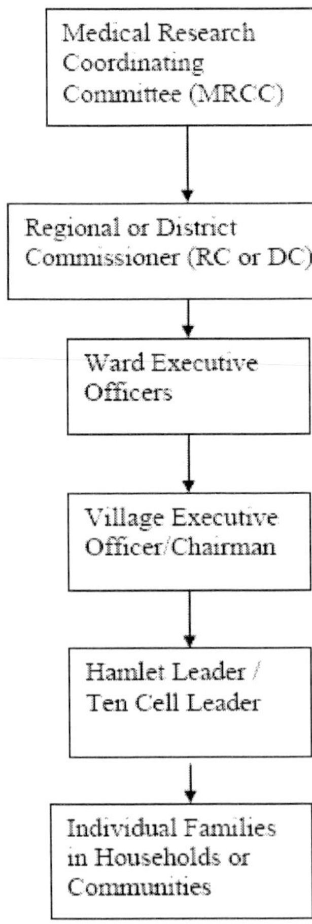

The Regional Commissioner (RC) and the District Commissioner (DC) is the boss of the Region or the district just like a governor in the US system. So he/she needs to be informed of what is going on in his/her own territory as in case anything goes wrong she/he will be answerable to the Presidents Office or the Prime Ministers Office. Furthermore, the RC's and

DC's these people are chairs of the regional or district security committees and are obliged to maintain peace in the region or district.

The ward executive secretaries are in-charge of the ward and they are answerable to the DC or RC in case anything goes wrong as they will be asked how did these people by pass the DC's or RC's office and why did you allow them. Definitely, without producing a letter from the DC's or RC's office you can not be allowed to enter the communities even if you have the ethical clearance certificate. You will always be told to start with any one of these 2 offices (DC's or RC's office) in that particular district or region.

The village executive secretaries or chairmen, these are in-charge of the villages and they are very close to their people in the communities as they also live in. For that matter they are usually listened to by the people as they know each other / them. They can mobilize the people quite easily in the villages and inform them any message or directives from above which need their joint implementation as villagers. Furthermore, it is the same people in the village who voted them in power so that they can lead them. So these village secretaries and chairmen are also answerable to the ward secretaries and even to the DC and / or RC.

The ten-cell leader is basically an elected local administrator who normally controls 10 households within a particular village. However, sometimes these ten cell–leaders control more than 10 households but the name actually means controlling ten households under normal circumstances or situations. The ten-cell leaders are very close to their people and as well their living in the same village or community and they are familiar with their people as they have a small control of 10 households or families. They are responsible for mobilizing their people in the 10 households and giving whatever information or message they may have been instructed by the DC, ward executive secretaries and / or the village secretary or chairmen. Thus, they are also answerable in the same fashion (i.e. village executive secretary, ward executive secretary and DC or RC).

Finally, a household represents people or family members living together, sharing and eating food from the same pot. The head of household is, if present is normally the father/man. However, in the absence of the father/man or for divorced and/or widowed women they play both roles as mother and father (head of household). All these people that is, the RC, DC, Ward executive secretaries, Village executive secretaries / chairmen, ten cell leaders, heads of households and community members in general in that particular district or region are responsible for keeping peace, respect to each other and harmony. Any violation of peace and harmony necessitates other governmental machinery to intervene such as the Police and the Tanzania People's Defence Forces.

National Bioethics Processes in Evaluating Biomedical Research into Country

A proposal is first submitted to the Secretariat of the MRCC/ COSTECH or IRBs of academic institutions like the Muhimbili University College of Health Sciences (MUCHS), The Kilimanjaro Christian Medical Centre (KCMC) and Bugando University College of Health Sciences (BUCHS). The MRCC in Tanzania is the top most review board which includes members from various from various higher learning institutions in the country,

Ministry of Health and the Private Sector. The document (4 copies of proposals) is sent to 3-5 reviewers and the process takes about 8-12 weeks (2-3 months) or more to have an approval or not depending on the reviewers' comments as it may require as well the research team to revise their proposal and resubmit it again before it gets ethically cleared by the MRCC/IRBs. Guidelines for proposal development are available and can be obtained on request from the secretariat of the MRCC in Tanzania. Furthermore, there is a fee of 300$ for foreigner researchers and 100$ for researchers from Tanzania institutions and or local collaborators per proposal and a similar fee of 100$ for Tanzanian students for review and obtaining ethical clearance. These amounts should be paid on the day of submitting the proposal for the first time.

The MRCC secretariat is under the National Institute for Medical Research (NIMR) and NIMR partly operates under the Ministry of Health though most of its activities are autonomous. Thus, NIMR is a parastatal institution which was established by the Act of the Parliament in October, 1979 and became operational a year later. Furthermore, by law of the country the institute (NIMR) is charged with carrying out, controlling, coordinating, registration, monitoring, evaluation and promoting health research in Tanzania. Therefore NIMR has important roles to play in ensuring alignment with the set research priorities of the country. The ethical clearance certificate issued by the MRCC bears two signatures; one of the NIMR-Director General and the Chief Medical Officer-Ministry of Health-Tanzania.

Monitoring of Biomedical Research

This is normally done or achieved in three (3) ways: (a). Research Process; (b). Monitoring of Ethics and (c). Human Rights and Evaluation and Outcomes.

Research Process

Here the research team and the MRCC/IRB monitor whether the research is carried out as proposed or planned or not. Furthermore, is the proposed study able to finish within the proposed time frame or not. Does the study need to change some of their methods while in the mid of the study or data collection and in this case researchers need to record and state clearly why they decided to change the methodology they previously stated they would use. Here a full description justifying why is needed to be put in writing.

Ethics and Human Rights

This is the most important aspect in Biomedical Research Monitoring. All ethical commitment and human rights need to be fulfilled. Researchers will be monitored to see whether they follow the ethical procedures described in their proposal and if violation is identified, the whole team can be sued or receive termination of their project/study before it reaches completion. Furthermore, the research team or PI needs to submit periodic reports to the MRCC/IRB committee which meets periodically to review such reports so that they can see whether the team is following the laid down ethical procedures as indicated in their proposal as well as according to the MRCC/IRBs set of ethical standards.

Evaluation and Outcomes

For international and local research the parameters for outcome of say the Intervention needs to be monitored and evaluated. They need all to be reported to MRCC/IRBs. Furthermore, a final report after the end of the study needs to be submitted as part of documentation and information or dissemination of the study results/findings.

Current and Future Status of Biomedical Research in the Country

My discussion will be based on my field experience (urban and rural/community based studies) and other observations made in Europe and United States Institutions and in particular research activities carried out at Johns Hopkins University-Baltimore-State of Maryland.

From my own research experience in Tanzanian, when subjects are enrolled as adult healthy volunteers say when carrying out pharmacokinetic or bioavailability studies these individuals after prior examination and screening of their health status to see if they qualify to be enrolled in the study and after consenting to participate in the study, they are usually given some incentives in a form of monetary to cover their fare and food. It should be born in mind that these people spend quite a significant number of hours ranging between 8-12 hrs depending on the design of the study and the different time intervals the researchers wants to obtain the various blood or urine samples, etc.

In most rural areas or communities of Tanzania, these rural people do not get or are not given any form of incentive as it is not ethically allowed by most ethical committees (MRCC/IRBs) in most institutions in the country. So the only benefits these people in the research communities or participating in the research will be general for the entire community or district and / or region if any which may arise from the research findings. For example if it may be for changing the drug policy; improving and increasing drug supply in that particular district etc. Thus, my personal feeling are that these rural people in rural communities who are usually involved in the study apart from the general benefits which they may get after a while as it is not easy to get immediate effects or benefits. These individuals are exploited to some extent as we do not give them any motivation or incentives despite taking and wasting their time (say 1-3 hours for interviews or focus group discussions) as they would have done other productive things or even resting with their families. So there is a need to reconsider afresh at least they should be considered to get a small incentive package for these rural people participating in the study in these communities. This small incentive can be in monetary form within the range say a minimum of 2,000-5,000 thousand Tanzanian shillings (equivalent to 2-5$) or more depending on the project. It should be born in mind that these people especially in rural communities are poor people, so such an incentive is a good motivation. However, the Principal Investigator(s)/researchers (PIs) is or are not supposed / allowed to (break the news) tell participants when recruiting them in the very beginning that you will get money or an incentive if you participate in the study as this will be unethical to influence them to agree to participate. We only need purely voluntary participation in the study without any motivation or threats or fear that one would loose or not get benefits from not participating.

Experience or my observation from Europe (Karolinska Institutet-Sweden) and United States (Johns Hopkins University-Baltimore, MD), all research advertisements I have seen on the notice boards consider subjects time to be wasted and for that matter it clearly states that if you are willing and once you qualify to be recruited you will be paid or refunded for both your fare and time which have taken / lost or invested on this project. Thus, they are given in terms of monetary form and sometimes even motivating them to come for their appointments and also to continue resisting or abstaining from using the drugs intended. This is for all studies and more common for studies involving substance of abuse (alcohol, tobacco/cigarette, marijuana smokers, crack cocaine, cocaine and heroin drug injectors).

Finally, through my research training career in rural communities of Tanzania, I experienced and swallowed all the blames and criticisms which were thrown on me by the study participants when you finish interviews or focus group discussions (FGDS) with them they expected to get some motivation or incentives. However, they were being surprised when I just thanked them and left, but could read their faces and sometimes they whispered and were suspicious that I have hidden or eaten their pay because nowadays every Tanzanian knows that research is money. Additionally, things or their suspicion was more worse if I was accompanied by my counterpart who was from Sweden (a White man or a *Mzungu in Swahili*) they thought a "*Mzungu*" had brought in more money for that or those studies which we were together as co-researchers.

Politically, Tanzania is a stable country from top to bottom and most of the political leaders are very understanding. However, as researchers this may be warning or caution that once you are in the communities don't involve yourself with politics. This is because Tanzania adopted multi-party system since 1997 and to date there are more than 16 parties. People in the communities live in peace, harmony and respect each other despite of their differences in political ideologies. So if you take these cautions you will get maximum support from the political leaders at all levels. Do just what has sent you there and not politics unless you are a politician which I believe you are not. Thus, in this way all villages or communities in Tanzania have this well established stability and people see each other as brothers, sisters, friends, neighbours and you name it. It is really a friendly country and very welcoming to live in whether it is urban or rural.

Conclusion and Future Entry Requirements into Communities

Need to consider new areas for conducting studies or research as some areas are over researched and subjects get tired or bored of being asked more less similar questions sometimes. This was observed several times during my research training in that particular district because during interaction they told me or commented that another group of researchers was here one or two weeks ago. Despite the fact that each research work has its own research questions and goals, but for rural people they may not be able to distinguish this.

Future research needs to consider all parameters cited in this document such establishing good public relations with the communities from leaders to the common man (people), always make sure you don't break promised appointments as you frustrate them because they leave all they are duties to wait for the research team. Furthermore, there is a need to consider some form of motivation as these rural people to do not have any income to supplement. So once you take their day off that means that particular day is gone and nothing will come in. This needs to be reconsidered from a humanistic point of view. We are getting data from them but they are not selling for us, so it is the matter of researchers to think about them (people or subjects being involved in the study).

Acknowledgements

I would like to thank Dr L. Mboera-Director-ICT and Mrs J. Ikingura all of NIMR, Dar-es-Salaam, Tanzania for assisting with some important literature which lead to the production of this. I am also thankful to Novartis Co Ltd -USA for facilitating this exercise as it sheds light to foreign researchers and readers what should be expected or what preparations need to be undertaken before or when requesting for biomedical research ethical clearance certificates for a particular country. To my employer; the Principal, Muhimbili University College of Health Sciences (MUCHS), Dar-es-Salaam-Tanzania, thanks for granting me study leave to attend courses 2005/06 at Johns Hopkins University-Bloomberg, School of Public Health, Baltimore-MD-21205-USA and this exposed me to Norvatis Coordinator while in U.S. who requested me to contribute a chapter on this for Tanzania.

References

Medical Research Coordinating Committee (MRCC) of the National Institute for Medical Research Tanzania.

National Bureau of Statistics-URT (2003). National Population and Housing Census General Report, 2002.

National Health Research Priorities for Tanzania: Proceedings of the workshop, 10th – 12th October, 2005. Arusha International Conference Centre-Tanzania.

Nsimba SED. (2003). Exploring malaria case management of under five children in households and public primary health care facilities in the Kibaha district, Tanzania. PhD thesis submitted to Karolinska Institutet-Stockholm, Sweden.

www.nimr.or.tz

www.tanz.gov

In: A Gateway to Biomedical Research in Africa
Editor: Takafira Mduluza, pp. 91-100
ISBN: 1-60021-444-4

Chapter IX

Conducting Biomedical Research in Malawi: What Researchers Need to Know

Paul Ndebele and Joseph Mfutso-Bengo
College of Medicine Research and Ethics Committee; Private Bag 360, Chichiri; Blantyre 3, Malawi; Tel +265 1 674377 Fax +265 1 674 740

General Information about Malawi

Malawi is located to the south of the Equator between latitudes 9 degrees 45 minutes and 17 degrees 16 minutes, and longitudes 33 degrees 35 minutes and is 900 Kilometers long and its width varies between 80 Kilometers and 160 Kilometers. Malawi has a total area of 118, 484 square Kilometers and a total population of about 9.9 million people according to 1998 census.[1] Malawi is bordered by Tanzania, Mozambique and Zambia and is endowed with fresh-surface water in the form of Lake Malawi and others. Lake Malawi stretches 568 Kilometers long and 16 kilometers wide, accounting for 20 percent of Malawi's total surface area. Malawi has a varied topography of mountains and rivers, which have a direct effect on the climate with temperatures averaging from 14.4 to 32 degrees Celsius and largely has two main seasons namely, cold-dry and hot-wet. The hot-wet season is from November to April. Rainfall varies from 725 millimeters to 2500 millimeters. English and Chichewa are the official languages, with several other languages being of greater importance regionally.

The vast majority of the population (86%) live in the rural areas, earning their living from the land and many areas in Malawi are regarded as food deficit resulting in severe poverty. With a gross national product (GNP) of 170$ per person per year, Malawi belongs to the 10-20 economically least developed countries in the world. The economy of Malawi is dependant on the price of tobacco and substantial economic assistance from the IMF, the World Bank, and individual donor nations. Agriculture accounts for over a third of the GNP, about 90 per cent of export earnings and approximately three-quarters of total employment.

Despite economic difficulties, Malawi has invested a lot in health. Malawi has a reasonably good network of health facilities and for 90% of births, mothers have received ante natal care from a trained health worker, 86% of children aged 12-23 months possess a health card indicating a high under-five clinic attendance and about 90% of children are reported to have been vaccinated against the most common illnesses. The maternal mortality rate of Malawi is estimated at 620/100 000 live births. Life expectancy at birth is below 44 years, largely because of high mortality amongst the children and attrition due to HIV/AIDS. From 1000 children born alive, 133 - 137 die before the age of 1 year and 211 - 234 before the age of 5 years. Malaria, tuberculosis and HIV AIDS present some major health challenges for Malawi. There are currently various interventions aimed at reducing the impact of these national challenges.[2]

Biomedical Research Oversight in Malawi

Most of the biomedical research in Malawi is conducted by the College of Medicine of the University of Malawi based in Blantyre as well as by non governmental organizations (NGOs) and other institutions mostly affiliated with external universities. Such research is meant to address the needs of the country and ranges from preventive, curative and health services delivery improvement studies. Research oversight in Malawi falls under the jurisdiction of the National Research Council of Malawi (NRCM which is a Department under the Office of the President and Cabinet. NRCM was established through a Presidential Decree on 30th March 1974 to be coordinate and promote research.[3] As a statutory body, NRCM provides the general framework and guidelines for the conduct of research in Malawi. In undertaking these functions, the Council is guided by constitutional principles such as those relating to freedom of access to information, academic freedom, freedom of expression and other related principles; and the National Science and Technology Policy which provides for all types of research (natural/basic and social sciences and the arts).

National Health Research Sciences Committee (NHSRC)

To execute the mandate of coordinating and promoting research, The NRCM established Procedures and Guidelines for Research Clearance in 1989. While the 1989 edition of these guidelines played an important role in creating a coordinated research environment, it was later noted with concern that under these guidelines, research clearance took long and researchers could not implement projects in time hence, in some instances they could lose funds earmarked for projects. To facilitate the speedy approval of research projects, some of the functions of the NRCM were, thus, decentralized to sectoral institutions from September 1993. The National Health Sciences Research Committee (NHSRC) was formed in 1993 as a sectoral committee of the National Research Council of Malawi (NRCM) to coordinate, regulate and promote research in the area of health sciences. With a delegated authority of the

Governing Council of the NRCM, the NHSRC was transferred to be directly serviced by the Research Unit of the Ministry of Health after designating the Ministry of Health to be secretariat of the NHSRC following decentralization of research clearance in Malawi by NRCM.

The Ministry of Health has a sectoral obligation and interest in seeing that regulations in health research are being adhered to by stakeholders. In this regard, NRCM, NHSRC COMREC and the Ministry of Health work hand in hand. The Ministry of Health is responsible for developing and reviewing health policy. Where such a policy deliberately needs to integrate issues of health research, the relevant bodies, in this case, the NRCM COMREC and NHSRC are brought on board to deal with the specific issues. The membership of NHSRC is by area of specialty and includes all major health specialties in Malawi. Membership also includes representatives of the lay public (1), social scientist (1) and a legal representative (1). Members from any institution in Malawi may be appointed. The NHSRC has sub-committees including a monitoring sub-committee for effective management of its affairs. NHSRC has provision for co-opting individuals outside NHSRC into its committee or sub-committees for specific and time limited assignments. Membership of NHSRC excludes research project directors as a way of managing conflict of interest. The Contact address of the NHSRC is as follows:

E-mail: doccentre@malawi.net
The Secretary
National Health Science Research Committee (NHSCRC)
Ministry of Health and Population
P.O. Box 30377 Capital City
Lilongwe 3, Malawi
Tel : +265 8 588 957 or +265 1789 400

College of Medicine Research Ethics Committee (COMREC)

In order to provide timely review and clearance of proposals from faculty members and students of the College of Medicine, the MOHP in 1996 authorized the formation of the College of Medicine Research Committee (COMRC) as a sub-committee of the NHSRC. Like the parent committee, the COMREC is also charged with the responsibility to monitor research projects it approves. COMREC operates from the College of Medicine of the University of Malawi. COMREC handles research conducted by investigators from COM and UNIMA constituent Colleges and their collaborators. COMREC refers all research of "national importance" to NHSRC for review and approval. According to its constitution, COMREC should be made up of 16 members representing the following: projects, BMS Division, Clinical Division (Obstetrics, Gynaecology and Surgical Departments), NHSRC, Kamuzu College of Nursing, Medicine and Paediatrics, Community Health Division, Pathology Division and 2 Lay community reps. The following 4 representatives are ex-officio; COM librarian, Medical Director QECH, Dean Postgraduate Studies and Research and a Bioethicist from the Bioethics and Research Unit. COMREC has a Secretariat which is well established and has all the necessary equipment such as computers, internet access, fax, laptop, projector, email etc. COMREC is housed in the Department of Hematology but there are future plans to move to the proposed Postgraduate Institute. COMREC is funded from the College of Medicine Budget and also from the $100 processing fee which was initiated in 2003.

Contact information:
e-mail: comrec@medcol.mw
The Secretary
College of Medicine Research and Ethics Committee
Private Bag 360
Chichiri
Blantyre 3, Malawi
Tel +265 1 674377 Fax +265 1 674 740

Relationship between NHSRC and COMREC

NHSRC and COMREC are the only two Ethics Review Committees in Malawi which are recognized by Government. They operate as sister committees acting within their properly defined jurisdictions and both report to NRCM. The NHSRC operates from within Ministry of Health and Population Head office (MoHP) in Lilongwe whilst the COMREC operates from the College of Medicine of the University of Malawi. NHRSC specifically deals with research being conducted by MoHP and other organizations and institutions which are not part of University of Malawi (UNIMA) as well as all research which is of national importance while COMREC handles research conducted by investigators from UNIMA constituent Colleges and their collaborators. The two Ethics Committees (i.e. COMREC and NHSRC) report to and are centrally monitored by NRCM. The protection of research participants and monitoring and evaluation of the approved research studies are primarily assured and done only by these two Ethics Review Committees in Malawi on behalf of NRCM. The NRCM remains the overall policy body.

To foster a functional relationship between these committees, the College of Medicine Research and Ethics Committee reviews protocols from members of College of Medicine and Kamuzu College of Nursing and their collaborators except protocols originating from studies of national interest. The National Health Sciences Research Committee reviews protocols from other researchers outside College of Medicine and Kamuzu College of Nursing and those referred to it by COMREC as being of national interest. Each committee has a member representing the other committee as a way of ensuring that the activities of the two committees are well coordinated.

Research of National Interest

Most health research being done in Malawi is generally of national interest. However, there are some studies that deserve particular attention because of their sensitive, political and safety implications. Studies covering the following areas are regarded as examples of "National Interest Studies" ; all vaccine trials; stem cell research; cloning research; all genetic studies; drug trials where patent issues are involved and where safety issues remain fully unknown; and national health surveys. All studies of "national interest" regardless of the origin of the protocol should be referred to the NHSRC who may form a standing committee for that specific project composed of members to be drawn on the basis of their expertise rather than which committee they come from. This ad hoc committee will monitor the project through to its conclusion. The project may be carried out in any geographical location as the committee sees fit. Such ad hoc committees include representatives from MoHP, NRCM and COMREC.

Research Guidelines Available in Malawi

The NHRC has issued the following National Guidelines copies of which may be available upon request to any of the ethics committees:

- Procedures and Guidelines for the conduct of Research in Malawi (NRCM, 2002)
- Procedures and Guidelines for Access and Collection of Genetic Resources in Malawi (NRCM, 2002).

The National AIDS Commission NAC which is based in the OPC has also issued the HIV/AIDS Research Strategy for Malawi for period 2005-2007. The research strategy seeks to address gaps in existing knowledge about AIDS in so as to inform policy, practice and HIV related interventions and thereby contribute towards the reduction of HIV transmission. The documents highlights the following guiding principles: political will and commitment, community participation, non-discrimination, de-stigmatization, promotion and protection of human rights, capacity building, partnerships and collaboration and the priority of HIV AIDS research (NAC, 2005). COMREC and NHSRC have also separately issued some guidelines on the submission of research proposals for review. These guidelines are available upon request.

Submission of Research Proposals to NHSRC and COMREC

Application forms are available from each committee upon request. Researchers are expected to complete the application form and to submit 15 copies of the complete application package to NHSRC and 4 copies to COMREC depending on which committee they intend to submit their proposal. COMREC charges a processing fee of USD$100 while NHSRC charges USD$150 for foreign researchers and MK500 (Malawi Kwacha) for Malawian Nationals. The application packet is expected to include the following; A letter from the respective head[s] of department(s) indicating that the research has the approval of the department(s) or institution(s), Copies of brief CVs for all investigators in the team. Research affiliations need a supporting letter from Head of Department. Both committees accept research proposals in the funding agency format as long as the proposal covers all the important sub headings such as objectives, purpose, justification, methodology and ethical considerations. Upon approval of projects, all applicants are required to pay to either NHSRC or COMRC a fee of 10% of the total budget indicated in the proposal. Detailed information on the requirements of each of the two Ethics Committees can be requested directly from either committee since these are revised from time to time.

Research Proposals by Foreigners

The Government of Malawi requires that foreign researchers intending to conduct research in this country should arrange for affiliation to Malawian research institutions to promote collaboration, improve on capacity building and exchange of expertise. Affiliation constitutes a contract between the foreign researchers and the affiliating institution. The latter will ensure that the researcher carries out investigations according to the agreed research methods and will monitor progress of the project. The ownership of the data generated from any research conducted in Malawi remains with the Malawi Government, irrespective of the source of finance of the research project. The researcher (either foreign or local) shall make available to Malawi at least four copies of all research findings obtained from the research work, two of which will be deposited at the National Archives and the NRCM besides the affiliating and other relevant sectoral institutions. Any foreign research project must negotiate a Memorandum of Understanding or an Agreement (as the case may be) which shall be mutually agreed and signed by the Project, the Affiliating Institution, Ministry of Health and the National Research Council of Malawi representing Government on research matters and acting within its mandate of co-ordination. This must be done before a project starts its operations in Malawi. For researchers intending to use funds from US agencies to conduct research, COMREC is registered with Office for Human Research Protections (OHRP) – Federal-wide assurance number is 00001395.

Clearance of Drug Trials

Drug trials normally fall into the category of studies of "national interest" depending on the disease the drugs are meant for. For all drug trials that are aimed at addressing diseases of National Importance, they have to be submitted to NHSRC for review and approval. For any drug trial to be done in Malawi, COMREC and NHSRC require that the drug under study must first be registered by the Pharmacy, Medicines and Poisons Board. The Pharmacy, Medicines and Pharmacy, Medicines and Poisons Board was established by Act 15 of 1988.

The Board's principal functions are registration of drugs, monitoring of drugs as well as registration and disciplining of pharmacy personnel. The proposals are only reviewed by one of the two committees and not by this Board. The Board is therefore not involved in research but only in the test drug for the purpose of licensing and monitoring. The Board also sanctions the importation of trial drugs into Malawi.

Enforcement Mechanisms for Violations of Human Subject Rules

For the purpose of addressing violations of human subject rules and enforcement of rules, Malawi relies on the common assault and battery laws. Issues related to Autonomy and justice are enshrined in the constitution of Malawi and in other pieces of legislation. There are other formal ways of dealing with enforcement and violations such as sanctions by Malawi Medical Association and Disciplinary Committees of the Constituent Colleges of UNIMA. Currently there are debates relating to the proposed Malawi Bioethics National Policy. The proposed policy includes compulsory training in Bioethics for all health personnel and courses in Research Ethics and Good Clinical Practice (GCP) for all members of research teams. The policy will also include some enforcement mechanisms. The NHSRC and the COMRC, respectively, are expected to thoroughly investigate all reported violations of protocol or ethical conduct. Where there is sufficient proof of violation, one or all of the following actions can be taken; the researcher may be asked to stop the study, the researcher may be asked to compensate study subjects where ethical violation took place and or the researcher may be taken to court.

Other Ethical or Regulatory Issues That Researchers Should Consider

Ethical review jurisdiction – research of National priority has to be reviewed by NHSRC and at times some problems due to the definition of national priority. Most health research being done in Malawi is generally of national interest. However, there are some studies that deserve particular attention because of their sensitive, political and safety implications. Studies covering the following areas are to be regarded as examples of "National Interest Studies": All vaccine trials; stem cell research; cloning research; all genetic studies; drug trials where patent issues are involved and where safety issues remain fully unknown; and national health surveys. All studies of "national interest" regardless of the origin of the protocol should be referred to the NHSRC who may form a standing committee for that specific project composed of members to be drawn on the basis of their expertise rather than which committee they come from. This ad hoc committee will monitor the project through to its conclusion. The project may be carried out in any geographical location as the committee sees fit. The ad hoc committees include representatives from MOH, NRCM and COMREC.

Research capacity building - Non-Malawian researchers are expected, as much as possible, to collaborate with local (Malawian) researchers to support efforts in capacity building. All research projects with non-Malawian Principal Investigator(s) should have Malawian Co-Principal Investigator(s) and the local counterpart should be involved from proposal development through to implementation. Research projects with no involvement of local personnel will rarely be approved.

Re-Use of Research Materials

Where a researcher intends to use research material collected for a different study, a new research proposal shall be required for review by the committee. In such cases the researcher shall be expected to specify how informed consent shall be obtained from owners of the materials.

Amendments to the Approved Protocol

Where it becomes necessary to make amendments to an approved proposal, the modified proposal should be submitted to the NHSRC for review and approval before implementation of the changes. Additional fees may be needed depending on the extent of the modification and the urgency for obtaining the approval.

Biological Specimen

The following statements explain the position of NCRM, NHSRC and COMREC on Biological specimens:

- Biological specimens should be collected only by a qualified individual. There should be proper handling of specimens to avoid wastage and recollection of the specimens;
- Researchers should refrain from collecting biological specimens that are not required to address their study objectives.
- Collection of specimens should be done only after the study subject or their representatives/caretakers have given informed consent.
- Collection of excess specimens from study subjects should be avoided.
    - Tests on biological specimens should only be as described in the approved protocol.
    - Specimens collected for a particular purpose should not be used for other purposes except where:
- Re-collection of specimens is not permitted except with approval from an ethics committee.

- The use of collected specimens for other uses may be considered where:
    - there is a clear national / individual benefit from re-using the specimens
    - there is need to confirm results due to technological or other problems
    - there is a way of obtaining informed consent from the subjects from whom specimens were collected.
- As much as possible analysis of specimens should be done within Malawi by local technicians / professional. In exceptional circumstances researchers may be allowed to export specimens especially where:
    - there is no technology available to conduct the desired tests
    - such technology can not be imported
    - further tests are necessary to confirm the results, and
    - quality control and validation are desired.
- It is recommended that a local (Malawian) technician or professional should accompany the specimens and participate in carrying out the tests where specimens are being exported to any other country, this is in line with capacity building.
- Principal Investigator should submit an application to the NHSRC or the COMRC for permission to export biological specimens. In the application, the researchers should explain why it is necessary to expert the specimens, how the specimen will be used, how long they will be kept and the name of the local technician / professional to carry out or participate in the testing.

Conducting Research in Malawi - The Future

It is envisaged that there are going to be some changes in research oversight in future. Presently various stakeholders are debating the Malawi Bioethics Policy which is aimed at ensuring the protection of Malawian patients and research subjects from abuse. The policy will make training in Research Ethics and GCP compulsory for all people working in the health delivery and health research areas. Courses on these topics are already being offered by the Bioethics Research Unit at the College of Medicine. These courses are aimed at undergraduate students, postgraduate students and other persons involved in the conduct of research such as researchers and ethics committee members. The College of Medicine is now also offering a Master of Public Health Degree with Specialty in Bioethics. The degree programme includes several modules in Research Ethics. These programmes are being offered through the Fogarty Research Ethics Training Program for Eastern and Southern Africa. It is hoped that through these various initiatives, the biomedical research environment in Malawi will reach greater heights.

References

National Statistical Office 1998 Malawi Population and Housing Census

The Malawi Government Gazette, 11th June, 1976, General Notice No. 398).

WHO (1998). World Health Report 1998

Helpful Documents

National Research Council of Malawi. (2002). Procedures and Guidelines for the conduct of Research in Malawi.

National Research Council of Malawi. (2002). Procedures and Guidelines for Access and Collection of Genetic Resources in Malawi.

The National AIDS Commission. (2005). HIV/AIDS Research Strategy for Malawi for period 2005-2007.

College of Medicine Research Ethics Committee Research Guidelines.

National Health Sciences Research Committee (NHSRC) Operational Guidelines

Helpful Websites

Government of Malawi: http://www.malawi.gov.mw/

Ministry of Health of Malawi: http://www.malawi.gov.mw/health/health.htm

Malawi Central Statistical Office: http://www.nso.malawi.net/

College of Medicine, University of Malawi : http://www.medcol.mw

In: A Gateway to Biomedical Research in Africa
Editor: Takafira Mduluza, pp. 101-108

ISBN: 1-60021-444-4

Chapter X

Republic of Cameroon

Bumah Violet Vakunseh, Ndeso Sylvester Atanga and Jane Akoachere
Faculty of Science, University of Buea, Box 63,
Buea, SW Province, Cameroon

Brief Introduction of the Status of Biomedical Research in Cameroon

Cameroon is a bilingual country made up of French and English speaking. Prior to independence in 1960 and 1961 respectively, Cameroon was divided into two parts: east and west Cameroon. The east was administered by the France while the west was governed by the British as part of Nigeria in conformity with a United Nations mandate. But in 1961, following a UN Plebiscite, the two parts united to form the Federal Republic of Cameroon. The federal system lasted up to 1972 when another referendum resulted into a Unity State. This was followed yet by another silent revolution in 1984 creating the Republic of Cameroon. Geographically, the country is in Central Africa situated deep into the Gulf of Guinea situated between the 2nd and 13th degrees latitudes north and the 9th and 16th degrees longitudes east. Triangular in shape, it is bordered in the north by Lake Tchad, in the east by Tchad, Central African Republic, Congo and Gabon, in the west by Nigeria and in the south west by the Atlantic Ocean. It covers a total surface area of 465 000 Km^2. The total population approximated from the 1997 census is 16 million inhabitants.

It is said that Cameroon is Africa in miniature. In effect, several types of natural relief contribute to the geographical diversity of the country. The south east, littoral and south west regions are covered by dense and humid tropical forests, the north is savanna and steppes while the west and northwest regions are less dense in forests but have rich volcanic soils that is good for agriculture. There are 230 ethnic groups in Cameroon.

From socioeconomic perspective, economic activities in Cameroon like in other developing countries is essentially agricultural based, and centered around production and export of raw materials of great variety (coffee, cocoa, bananas, cotton, rubber, timber, oil, palm oil, etc). The country enjoyed an economic boom in the seventies and mid eighties, but for the past two decades it has been experiencing serious economic crisis that has led to extreme poverty and low standard of living for the majority of its citizens. It is currently rated as a heavily indebted poor country by the IMF and World Bank. It is however expected that attainment of the heavily indebted poor country initiative completion point will reverse this dismal economic picture for the benefit and welfare of the citizens of this natural and human resources endowed country.

From a health point of view, the country counts 1 bed for 400 inhabitants and 1 doctor for 11 500 inhabitants. The country is self-sufficient in food although for geographic reasons some parts like the north frequently experience food shortages due to prolonged dry seasons resulting in drought. Infant mortality and maternal mortality remain high (110/1000 and 450/100,000 respectively). Life expectancy which was 57 years on the average in 1991 (Jeune Afrique Economie, 2000), is now put at 53 in 2005.

At the moment is uncertain as to whether biomedical research in Cameroon is organized. As such, it is equally not clear as to what the research priorities for the country are because there is no sound research policy. But existing evidence shows that for development purposes, two specific areas of research have been emphasized and encouraged in Cameroon. The two areas are Agriculture and Health Research. In both cases, research has been centered around the application of biotechnology in improving agricultural production and its application in the diagnosis and treatment of some endemic diseases. In this write up, we will only focus on Health Research in Cameroon. As in Tanzania, two broad categories of health research are identified: Health Systems Research and Biomedical Research. These two categories will be further divided into two distinct but related categories and prioritized as high, medium or low.

I. Biomedical Research: High priority is accorded to the following topics: communicable diseases of public health importance (malaria, trypanosomiasis, onchocerciasis, leprosy/tuberculosis); medium priority include disease control, non-communicable diseases, nutritional deficiencies; low priority include environmental health impact assessment and traditional medicine.
II. Health Systems Research: this include High priority diseases such as HIV/AIDS; reproductive health and health financing; medium priority include community access to essential drugs, health policy and low priority include community participation and donor participation in health financing (MOH, 1995).

It is important to note that government sponsored biomedical research is limited to HIV/AIDS because of its serious consequences and to a lesser extent research on ways to prevent and control the big endemic diseases. The first HIV/AIDS research center for Cameroon and Central Africa was inaugurated in February in Yaoundé 2006 by the first lady of Cameroon. Other types of research are carried out by individual scientists in their various laboratories for their own interests and with little or no coordination from the government. It is generally difficult to know who is doing what, why, and for who in biomedical research in Cameroon.

The Study Population and Communities Available for Biomedical Studies

Cameroonians are very friendly and kind people be it urban or rural communities and have been found to readily cooperate and participate in biomedical research studies as long as administrative authorities have approved such studies.

As in some other African countries, some communities may be suspicious of research that involves mostly foreigners. This belief is true of most African societies or most societies for that matter. It will of course be impossible to gain access to communities for the purposes of biomedical research or other activities if one were not from that area. For one thing, especially in Cameroon, the language barrier will be a problem. As mentioned above, there are 230 ethnic groups in Cameroon; all speaking various dialects or native languages. From a national perspective, you must speak the two official languages (French and English) and Pidgin English if you want to communicate with people all over the national territory. Consequently, biomedical research teams led by foreign experts or nationals must include research assistants preferably from the research setting who can communicate in the local dialect. If not the research idea whether beneficial or not to the population will be viewed with suspicion and disbelief leading to outright rejection.

Several cases have been reported in which medical teams leading Anti-Polio Vaccination Campaigns have been chased out from some villages in Cameroon because they believed the real purpose of the campaign was to sterilize their teenage girls thus rendering them barren for life. Therefore any biomedical research must be authorized officially and informed consent of the local population obtained through their traditional rulers. Our experience is that once the health authorities, local administrative and traditional rulers give their consent, it is usually very easy to enter most communities in Cameroon for the purposes of conducting biomedical research. Once the administrative authorities are informed, it is assumed that they in turn inform their subjects. It is therefore unnecessary to get everyone's consent before a study is conducted.

Cameroonians are neither rich nor poor. The country is endowed with abundant natural and human resources. The country is self-sufficient in food as mentioned earlier except for the northern territories where prolonged dry seasons may lead to occasional drought and food shortages. In fact the problem in Cameroon is one of equity in food distribution rather than shortage. Food is available in the south of the country in abundance, but due to poor road network, the excess food cannot be transported to those in need in the north.

More than 70% of the population live in the rural area and most are involved in small scale farming for both subsistence and income generation for foreign exchange earnings. Agriculture constitutes the mainstay of the economy. A good proportion of food produced is exported to neighboring countries of Gabon, Congo, Equatorial Guinea, Tchad and Central African Republic. Cameroon is considered the breadbasket in Central African Region.

Cameroon has been a peaceful and stable country since independence in 1961. Although it was ruled by a monolithic party structure for over 40 years, it adopted a multi-party democracy in 1992. There are over 110 political parties in Cameroon. The country is divided into 10 Administrative Provinces, 58 Divisions and over 800 districts for purposes of facilitating political and administrative governance. There are two main cities in Cameroon;

Yaoundé, the political capital city of over 1 million people, and the port city of Douala with approximately 2 million people. Douala is also the economic capital of the country. Politically, the country has adopted a Presidential system of government but government business is headed by a Prime Minister appointed by the Head of state from the ruling party (Jueune Afrique Economie, 2000).

Procedure for Accessing Communities for Purposes of Biomedical Research

There is no national organ charged with issuing ethical clearance for biomedical research in Cameroon. As such, there is no talk of ethical clearance for research but authorization to conduct research. However, two government ministries: The Ministry of Public Health (MOH) and the Ministry of Technical Scientific Research (MINREST) can authorize biomedical research in Cameroon. The MOH will issue authorizations for non-university based researchers while MINREST issues authorizations for university academics.

Depending on the region the study is to be conducted; the governmental structure having jurisdiction over the local administrative service in the area where the study is to be conducted; it is recommended that the authorization be obtained from that service's central administration from the national capital, Yaoundé. For example, if a non-university based researcher wanted to study the impact of using impregnated mosquito bed-nets on malaria transmission in Yaoundé or any other city in Cameroon, the researcher will need to obtain permission for the study from the Minister of Health in Cameroon. A University-based researcher will obtain authorization from MINREST for a similar or related study.

However, it is also noted that depending on the subject being considered for investigation and to hasten the authorization process, some Provincial Delegate of Public Health (PDPH) have been given the mandate to authorize researchers; university or non-university based to conduct some biomedical research in areas within their jurisdiction.

Once the authorization is issued, the next step is to get permission to enter the district where you want to carry out the research. Whether the researcher is a Cameroonian or foreigner, the first administrative person to contact is the Governor of the Province (GP). The Senior Divisional Officer (SDO) and the Sud-divisional Officer (DO) assist the Governor. In the absence of the Governor, the SDO can permit researchers to enter the province, the division or sub-district for purposes of conducting biomedical research.

In the governor's office, the research team introduces itself; this is then followed by, a brief explanation of the nature and purpose of the research and finally, a copy of the signed authorization of the research from MOH or MINREST is given to the governor for his information. In most cases, governors are most welcoming once they have been briefed on the purpose of the research. They may comment or ask a few questions about the research, but in general they laud the initiative and encourage the team to carry out their study to a successful conclusion for the benefit of the people in the area and Cameroon.

From the governor, the next person to be consulted is the Provincial Delegate of Public Health (PDPH). The PDPH is briefed on the research and then shown the letter of authorization. The delegate may give the team his blessing or he may issue another letter introducing the research team to the District Medical Officer (DMO). From the PDPH, the next port of call is the DMO's office. DMOs are very cooperative. They usually brief research teams on the organization and functioning of district health services under their jurisdiction. From the DMOs, the team can either go directly to the village or they may visit the Health Centre (HC) for courtesy purposes. The HC is the population's first point of contact with the health system. It is usually good to visit and brief the Health Centre Chiefs (HCCs) on the purpose of the research. HCCs play an important role in that they can facilitate contact with the traditional rulers or chiefs as they are called, of the village in which the health center is located. Traditional rulers are very powerful and influential administrators in their areas. Contact with them is not direct except through an intermediary. There is also the Health Centre Committee, a co-management structure that is very important in mobilizing the community for health action. Some members of this committee may be co-opted to assist the team in entering the village. For this particular reason, it is important to visit the health center.

The Traditional rulers in Cameroon are very warm and friendly to visitors especially those that come to carry out duties that can improve the welfare of their subjects. They are very instrumental in ensuring access to households in their villages. Without their consent, it is impossible to carry out any study in the village. If a research team is found carrying out a study in the village without the permission of the Chief, the entire team can be mal-handled whether or not a Cameroonian is in the team. The chief is administratively assisted by Quarter Heads (QH) who are direct representatives in the various quarters of the village. They report directly to him. Quarter heads usually know the heads of the various households in the village. They also know most of the very active young men in the village. They can therefore be very instrumental in helping research teams in identifying guides that can take teams to various houses especially in studies that require house to house survey interviews (Ndeso et al. 2000). Such guides can also act as interpreters especially in homes where there are old folks who can neither speak, French, English nor pidgin.

A household is made up of both members of the nuclear and extended family living together, sharing and eating from the same pot. The head of the household is usually the husband. But when there is no husband as it happens when he is dead, or when there is a divorce, the woman plays the role of father and mother and as head of household. The Governor, the Senior Divisional Officer, the Sub-Divisional Officer, the Traditional ruler or King, the Village Chief and Quarter Heads and Heads of Households are responsible for maintaining peace and harmony and mutual respect in the province.

National Bioethics Processes in Evaluating Biomedical Research in Cameroon

As mentioned before, there is no particular organ in Cameroon that is charged with the screening and evaluation of research proposals that involves human subjects in order to make sure that they conform to required ethical standards. If at all this body exists, it does not take its work seriously. The review process is not as rigorous as in Tanzania where formal ethical clearance is necessary. In applying for authorization to carry out biomedical research, one may submit a copy of the proposal, or a summarized version of it to MOH, MINREST or the PDPH. It might take a few weeks or a month to obtain the written authorization. The time it takes from the moment a proposal is deposited to when an authorization is issued is not a function of the proposal being subjected to some peer or board review, but that the responsible official may just be taking time to understand what the study is all about. There is an Institute of Medical and Medicinal Plants Research (IMPM) which is one of the research institutes of MINREST, but it is hard to tell whether it carries out ethical clearance as is done by the MRCC/IRBs of Tanzania. We have conducted biomedical research that to the best of our recollection did not have to be ethically cleared by this body. To conclude, it can be stated that the absence of bureaucratic red-tape makes it easier to obtain ethical clearance for biomedical and other types of research in Cameroon (Ndeso et al. 2000).

Monitoring of Biomedical Research

The investigators themselves and the sponsors of the study monitor the research process. Organizations that fund research work with investigators to ensure that the work is completed on time. Data collection methods may change in the middle of the research but one does not necessarily have to justify in writing the need for such change. It is the prerogative of the investigators to ensure that ethical standards as stated in the proposal are maintained throughout the study period. This however does not mean that investigators have the right to resort to unethical means just for the purpose of bringing their study to a successful conclusion. It is generally assumed that researchers are aware of the ethical implications of their research and that they will take every precaution to protect the subjects and the community they are studying. There is of course a danger with such flexibility. In the event that ethics and human rights of subjects are not safeguarded as sometimes happens, the whole team may be brutally attacked by mob action and the project prematurely terminated. This may happen before the forces of law intervene.

Evaluation and outcomes: We are not aware of the existence of any parameters for the evaluation of intervention outcomes. Incidences of possible adverse effects of certain interventions based on HIV/AIDS research findings have been reported, but not scientifically documented in manner that may require that the investigator to be called to justify his claims of therapeutic efficacy. The absence of a systematic way of disseminating the results of research findings to the public or to a duly constituted scientific body makes it difficult to monitor and evaluate research outcomes. Results of research findings are usually the property of the principal investigator or those funding the research. In most cases, such funding agencies are outside the country, and this presents a very serious obstacle in monitoring interventions that may have emerged from such studies.

Current and Future Status of Biomedical Research in Cameroon

Biomedical research needs to be more organized and better managed in Cameroon. While the major obstacle may be funding, the review has shown that obtaining authorization to conduct biomedical research is not difficult in Cameroon. A research body that deals with the whole issue of ethical clearance needs to be put in place. The Institute of Medical and Medical Plants Research (IMPM) can play this vital role. There is no need for two ministries to perform this very sensitive task of granting ethical clearance for research. This same goes for monitoring of the research process in the field and evaluation of outcomes and impacts of such research. IMPM has qualified scientists and experience in conducting biomedical research. It can therefore perform these important functions.

The issue of incentives for rural people participating in research is a good one, but we need to be very cautious as to how we go about encouraging subjects that participate in studies. For clinical trials to test the efficacy of a new drug, we agree that there should be some incentives especially when we consider the risks that untoward side effects could have on the individuals participating in such trials. We also agree that it should be a matter of policy that such individuals be compensated. We have traditionally given some token amount to thank research guides and interpreters who assist us in house-holds surveys in either rural or urban research settings. Should this amount be standardized? We have no answers at the moment, but we think it should depend on the site and the socio-economic status of the community. However, we have found that participants; be they research assistants or subjects participating in both urban and rural studies, perform better when some incentive is given. If there is none, the research team risk being abandoned especially when they have advance knowledge that the research is being financed from external sources.

We cannot and should not aim at compensating people in general for participating in one way or another in a study. The danger is making compensation or incentives a prerequisite for taking part in field studies. Another danger is the cost involved in such a policy. What needs to be done is to evaluate each situation on its own merit and then determine what will be the ideal incentive package. This should be the way forward.

References

DHS (1991) National Demographic and Health Survey Report. Published by Macro International Inc. Pp 5-30

Jeune Afrique Economie (2000) Cameroun: Politique, Economie, Industrie, Commerce et Development. Published by Paschal Tala

MOH (1995*. The Reorientation of Primary Health Care in Cameroon: Development Discussion Papers Series: Harvard Institute for International Development

Ndeso S.A., Leonard D.K. (2000) Choice and Quality of Health Care Providers in Rural Cameroon: In Changing Markets for Human and Veterinary Medicine Services in Africa: New Institutional Issues: Published by MACMILLAN UK Ltd. Pp 120.144

In: A Gateway to Biomedical Research in Africa
Editor: Takafira Mduluza, pp. 109-116
ISBN: 1-60021-444-4

Chapter XI

Biomedical Research in South Africa

Aceme Nyika[*]

South Africa, measuring 1,219,912 sq km, is located at the southern tip of the continent of Africa. Mozambique, Swaziland, Zimbabwe, Botswana and Namibia are the five countries that share borders with South Africa while Lesotho is completely surrounded by South Africa. According to the 2001 South African census report, the population is 44.8 million. About 30% of the population is aged 0-14 years, 65% is 15-64 years old and 5% is aged 65 and over. The median age is 23.98 years for the total population, 23.12 years for males and 24.86 years for females. Life expectancy at birth rose from 53 years in 1970 to 62 years in 1990 and then drastically dropped to 47 years as of 2004, mainly due to the HIV/AIDS epidemic. Unemployment rate is estimated to be 25.2%.

Current and Future Status of Biomedical Research in the Country

For a developing country that has recently emerged from political conflict, South Africa has made considerable progress in terms of promotion of biomedical research as reflected by firstly growing government support for research and secondly increasing number of companies and organizations producing a wide range of novel biomedical products. The South African government, through the Department of Science and Technology (DST), developed the National Biotechnology Strategy in 2001, which highlighted the potential contribution of biotechnological research to improvement of health, food security, environmental conservation and economic growth. Similarly, the Health Research Policy in South Africa adopted by the Ministry of Health in 2001 seeks to "provide an enabling framework for the conduct of research that improves human health and wellbeing in South

[*] Aceme Nyika; Centre for the AIDS Programme of Research in South Africa (CAPRISA), Nelson R. Mandela School of Medicine; P. Bag X7 Congella 4013; University of KwaZulu-Natal; 719 Umbilo Road; Durban 4001, South Africa. Email: nyikaa@yahoo.com

Africa". Overall, the policy framework is aimed at maximizing use of research output from academic research institutions to address real national health and developmental needs in partnership with the private sector.

The main government-created organizations involved in research capacity building, technology transfer, funding research and development (RandD) and regulatory and legal issues include the Council for Scientific and Industrial Research (CSIR), Medical Research Council (MRC), National Research Foundation (NRF), Godisa Trust, Technology and Human Resource for Industry Programme (THRIP) and South African National Bioinformatics Institute (SANBI). The NRF, which was created by an act of parliament (NRF Act of 1999), promotes research through provision of research funds, human resource development and improvement of research facilities. In an effort to integrate knowledge, innovation and productivity into an economic base for the development of the country, the NRF fosters strategic partnerships with local private and public organizations as well as with international collaborators. The implementation of national research policies rely to a large extent on the role played by a total of 21 universities in the 9 provinces of South Africa.

A national biotech survey conducted in 2003 showed that there are about 106 South African companies whose core business involves biomedical products or services. Out of these companies, 29% are direct spin-offs from research groups. Although there are a number of multinational companies in South Africa, including Pfizer SA, Novartis SA, GlaxoSmithkline SA and Merck SA, their contribution in terms of funding biomedical research is relatively minimal. Also relatively low is the level of venture capital investment in biomedical research. Thus the positive support and political will that the South African government has demonstrated so far is very crucial for the future of research and development in general and biomedical research in particular.

Disease Burden and Research in South Africa

Biomedical research is one of the tools that are being used to combat infectious and non-communicable diseases that are causing high rates of morbidity and mortality. For instance, it is estimated that by the end of 2004, 4 million South Africans were living with HIV/AIDS. According to the "3 by 5" progress report, only 7% of the people needing antiretroviral (ARV) treatment in South Africa were receiving ARVs by June 2004. Consequently, the country has intensified efforts to combat the epidemic through preventive and treatment programs. Preventive programs include behavioral change interventions and biomedical studies aimed at developing protective vaccines. The South African AIDS Vaccine Initiative (SAAVI) is playing a leading role in conducting research that is crucial for the development of vaccines for HIV/AIDS. The first human trials of a genetic HIV vaccine derived from HIV-1 subtype C, which is the most prevalent in Africa, were initiated by SAAVI in South Africa in 2003. SAAVI collaborates with other global players in HIV research such as the NIH-funded HIV Vaccines Trials Network (HVTN), the African AIDS Vaccine Programme (AAVP) and the European Developing Country Trial Platform (EDCTP).

Other major diseases that beset South Africa include tuberculosis, malaria, respiratory and diarrhoeal infections, cancers, diabetes, cardiovascular and kidney diseases. Research being conducted on these diseases range from molecular basis of illness to diagnostics and treatment methods. Examples of South African products or services derived from such biomedical research are Rhesus antigen typing kit for blood typing, diagnostic pathological services for various diseases and recombinant human erythropoietin for anaemia. However, in light of the magnitude of the disease burden of South Africa, biomedical research needs to be scaled up and international collaboration could go a long way in complementing government efforts.

Position of South Africa on Human Cloning

South African legislation and ethical guidelines prohibit gene modification of human germ line (gametes or zygotes) and human tissue cloning for reproductive purposes. Cloning of human stem cells is permitted only for therapeutic purposes and under stringent regulations. Cadaveric fetal tissue and surplus embryos from infertility treatments are the only permissible sources of stem cells for such therapeutic cloning. In addition, it is illegal to sell or buy embryos and cadaveric fetal tissue. Some ordinary people may be uncomfortable with the idea of giving human samples such as blood for research purposes. This is due to cultural or religious beliefs that view human tissue as being too sacred to be given away to other people, especially of different cultural background. It is therefore important to plan for community sensitization programs ahead of such research projects in order to ensure comprehension of objectives of the intended project and cultivate a sense of mutual trust between the researchers and the target communities.

The Study Population and Communities Available for Biomedical Studies

The country attained its independence from Dutch settlers (Boers) in 1994, ending a racist policy of apartheid. However, the consequences of the historical separate development of races are still reflected in the South African population in terms of socio-demographic factors such as level of education, employment opportunities, urban and rural settlements and health status. By 2002 about 98% of urbanites were using safe drinking water and 86% had access to adequate sanitation facilities in the urban areas while 73% and 44% of the rural population were using safe drinking water and had access to sanitation facilities, respectively. The adult literacy rate is 86%.

Due to the prevailing socio-economic factors, the poor communities in high-density locations or informal settlements in urban areas and in the rural areas carry the larger burden of ill health. Consequently, members of the poor communities constitute a major proportion of participants in health research and/or intervention programs. It is therefore important to have an understanding of the composition and cultural practices of the South African population before embarking on research in the country.

Composition of the Population

Independence ushered in an era of hope for a peaceful, democratic and prosperous new South Africa, popularly called the 'rainbow nation'. However, ethnic, language and religious differences have a bearing on people's health belief models and their response to ill health. It is therefore imperative that health researchers are sensitive to these differences when they design and implement biomedical studies. The composition of the South African population in terms of ethnicity, languages and religious affiliation is given in Table 1.

Table 1. Ethnic groups, languages and religions according to 2001 census

Ethnic Groups	Languages	Religions
Black African (79%)	IsiZulu (23.8%)	Zion Christian (11.1%)
White (9.6%)	IsiXhosa (17.6%)	Pentecostal/Charismatic (8.2%)
Colored (8.9%)	Afrikaans (13.3%)	Catholic (7.1%)
Indian/Asian (2.5%)	Sepedi (9.4%)	Methodist (6.8%)
	English (8.2%)	Dutch Reformed (6.7%)
	Setswana (8.2%)	Anglican (3.8%)
	Sesotho (7.9%)	Other Christian (36%)
	Xitsonga (4.4%)	Islam (1.5%)
	Other (7.2%)	Other (2.3%)
		None (15.1%)
		Unspecified (1.4%)

Heath Care Beliefs and Practices

It is estimated that about 80% of the South African population believe in and make use of traditional medicine. Thus in the health belief model of the majority of people, traditional medicine is the first port of call since illness is usually associated with witchcraft, ancestral spirits and/or sorcery. Most people believe that biomedical treatment cannot be effective unless the fundamental causes that are believed to be of spiritual nature are taken care of by traditional healers and/or spirit mediums. The traditional healers prescribe remedial measures that may include crude mixtures of herbs, rituals, or sacrifices, singly or in combination. There are no standard criteria for a person to be proclaimed a traditional healer and start practicing, which has recently compelled the South African government to initiate legislative process to try and control the practice of traditional medicine in the country.

As for religious people, they may choose to resort to their religious healers for healing prayers. Although some religions do not necessarily approve of African traditional medicine, many followers of such religions secretly consult traditional healers because it is a traditional practice that has been passed on from generation to generation. Thus there are basically three treatment choices, namely biomedicine, traditional medicine, and religious healing. Typically, biomedical treatment is sought for emergencies and acute ailments such as fever, diarrhea and injuries whereas for chronic illnesses, which are generally associated with supernatural causative agents, traditional or religious healers are consulted to address the causes while biomedical medicines may be sought to treat the biological illness.

Food and Child Rearing Practices

Since South Africa has a very heterogeneous population, made up of Blacks, Coloureds, Indians and Whites of different economic status within and between races, diet varies enormously. Broadly, the majority of urban dwellers are better off than the rural folks, hence they have adapted to a diet with rice and beef, mutton, pork, chicken or fish as relish. The rural populations typically eat thick porridge made from maize meal with vegetables as the main relish but occasionally mixed with meat. The common fruits available include mangoes, bananas, oranges and apples. Alcohol and drug abuse have been on the increase since independence, especially among teenagers.

Child rearing practices may vary slightly depending on ethnicity and socio-economic status. However, there are three main stages in the process of raising a child. The first stage is the infancy stage, which is characterized by nursing of the infants by the mothers, involving breastfeeding either exclusively or in combination with other liquids and solid foods. Typically, weaning occurs at about 2 years or when the mother becomes pregnant. Although immunization coverage rates against tuberculosis (BCG), diphtheria pertussis and tetanus (DPT1 and DPT2) poliomyelitis (Poio3) measles, hepatitis B (HepB3) and haemophilus influenzae have risen to over 90%, basic heath indicators such as the under 5 and infant (under 1) mortality rates have increased from 60 to 67 and 45 to 54 per 1000 live births, respectively, from 1990 to 2004. During the initial stage of child rearing, the mother can make decisions on issues such as taking the child to the clinic for immunizations or for minor illnesses without necessarily seeking permission from the father or in-laws. However, for proxy consent for a minor to participate in a research study, especially involving invasive procedures or several visits to the health center, permission has to be sought from the father and/or in-laws.

The second stage, from the age of about 7 years, the father and other members of the extended family begin to play an increasing role in the upbringing of the child. Generally, female relatives guide girls while male relatives take care of boys. The third stage begins at puberty, when the nurturing of the children into young adults begins. In some cultures boys undergo an 'initiation' ceremony that in some cases involves circumcision. For girls this stage typically starts when they begin to menstruate, and may involve regular virginity tests in some communities. Although the legal age of majority is 18 years in South Africa, teenagers younger than 18 who get married or cohabitate with their partners may be considered to be emancipated minors who can give informed consent to participate in health research, provided the potential risks are minimal and waiver of proxy consent requirement has been given by an appropriate ethics committee.

National Bioethics Processes in Evaluating Biomedical Research

Although the Medical Research Council (MRC) is a national board that formulates ethical guidelines for South Africa, national or international collaborative research does not necessarily have to be approved by the MRC. Institutional review boards that are properly constituted as per the national laws and guidelines can approve biomedical research proposals to be conducted in the country. The MRC formulates research ethics guidelines for use by ethics committees and researchers. In order to ensure compliance with set national ethical standards and regulations, a central body, the National Research Ethics Council (NHREC), was legislated by an act of parliament (National Health Act No. 61 of 2003) to register and monitor ethics committees. In addition to the regulatory mandate, the NHREC should also be involved in capacity building and public awareness programs.

Monitoring of Biomedical Research

Ethics committees are responsible for monitoring all research projects that they approve. Adverse events must be reported to the relevant ethics committee as soon as they occur. Any deviations from the approved protocol must be authorized by the same committee. In addition, researchers are required to renew their study approvals annually by submitting progress reports to the responsible ethics committee. Any payments for participation in a research project have to be approved by the ethics committees. Generally, permissible justification for such payments is considerable loss of income due to participation in a study.

The South African Medicines Control Council (MCC) is mandated to register drugs that have been proven to be safe and efficacious through properly designed scientific studies. Thus the MCC complements the role of ethics committees of protecting the welfare of people by ensuring the safety and efficacy of drugs that go onto the market.

Entry Procedure into Community for Biomedical Research

Whenever and wherever possible permission to enter into a community should be sought from community leaders. Local district and/or provincial heath officials should also be informed. However such permission to enter into a community should not be regarded as community consent on behalf of ordinary members of the community. South African ethical guidelines and regulations require first person informed consent or proxy consent (in the case of minors) to be obtained from research participants.

At the family level, men are generally the head of households and have to be consulted before family members can decide to take part in heath research. Family consultations involve spouses for married people while unmarried people consult close relatives. Male prospective participants usually consult such relatives as brothers, uncles and friends while females consult their sisters, aunts or friends. Thus it is important to allow time for such consultations during the process of recruitment and obtaining informed consent.

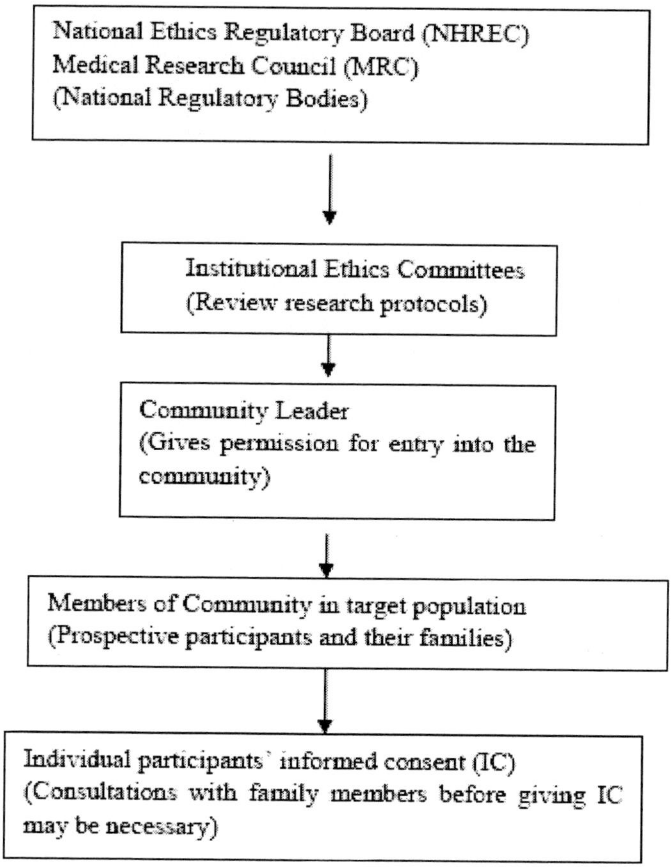

Diagram 1. Entry processes into target communities for biomedical research in South Africa.

Conclusion

Peaceful political and promising economic conditions prevailing in South Africa are conducive to the conduction of biomedical research. The South African government recognizes the critical role of research in the development of the country hence the National Research Foundation, NRF, was set up through which the state channels resources to support research in priority areas that include human health. In order to bring in additional resources and expertise, there is political will to encourage international collaboration. Although there has been an increase in the amount of health research conducted in South Africa since independence, it still does not match the disease burden of the country. Thus there is need for more research in the health sector if the health needs of the local population are to be effectively tackled. Because of its good technological infrastructure and policies, South

Africa has great potential to increase biomedical research leading to patentable and commercializable novel products that could go a long way in improving the health status of its people and the economy.

References

Bradshaw D, Groenewald P, Laubscher R et al. Initial burden of disease estimates for South Africa, 2000. S. Afr. Med. J. 2003; 93: 682-688.

Bradshaw D, Nannan N, Groenewald P. et al. Provincial mortality in South Africa, 2000: priority-setting for now and a benchmark for the future. S Afr Med 2005; 95(7): 496-503.

Council for Scientific and Industrial Research (CSIR). http://www.csir.co.za/plsql/ptl0002/ptl0002_pge001_home

Crabb C. 10 years after apartheid. Realigning South Africa's Science. Science. 2004; 304(5669): 376-7.

Dhai A, Moodley J, McQuoid-Mason DJ and Rodeck C. Ethical and legal controversies in cloning for biomedical research--a South African perspective. S Afr Med J 2004; 94(11): 906- 909.

Godisa Trust http://www.godisa.net/

Isaakidis P, Swingler GH, Pienaar E, Volmink J and Ioannidis JPA. Relation between burden of disease and randomized evidence in sub-Saharan Africa: survey of research. BMJ 2002, 324:1-5.

Katz I. Kidney and kidney related chronic diseases in South Africa and chronic disease intervention program experiences. Adv Chronic Kidney Dis. 2005; 12(1):14-21.

Muller M and Henschel T. National Biotech Survey 2003. Idea to industry. Prepared for eGoli BIO and Department of Science and Technology, South Africa. http://www.egoli.co.za/pages/biotech_survey.pdf

National Research Foundation (NRF) http://www.nrf.ac.za/

Progress on Global access to HIV antiretroviral therapy: An update on “3 by 5” June 2005. Report to Parliament 2005: Alcohol and drug abuse research unit. http://www.mrc.ac.za/adarg/parliament2005.htm

Sidey P. South Africa to regulate healers. BMJ 2004; 329:758.

Siegfried N, Clarke M and Volmink J. Randomised controlled trials in Africa of HIV and AIDS: descriptive study and spatial distribution. BMJ 2005, 331:1-6.

South African AIDS Vaccine Initiative http://www.saavi.org.za/

South African Medical Research Council (MRC) http://www.mrc.ac.za/

South African National Bioinformatics Institute http://www.sanbi.ac.za/

South African Universities http://www.statssa.gov.za/census01/html/C2001PrimTables.asp

Steyn M. Muti murders from South Africa: A case report. Forensic Science International 2005; 151:279-287.

Technology and Human Resource for Industry Programme http://www.nrf.ac.za/thrip/

http://www.cia.gov/cia/publications/factbook/geos/sf.html#Issues

In: A Gateway to Biomedical Research in Africa
Editor: Takafira Mduluza, pp. 117-130
ISBN: 1-60021-444-4

Chapter XII

Biomedical Research in Zimbabwe

Nicholas Midzi and Takafira Mduluza*

Brief Introduction on Country Biomedical Research Status

Zimbabwe is a land locked country with an estimated surface area of about 390 759 Km^2. It borders with Mozambique that lies in the East, South Africa in the South, Botswana in the West and Zambia that lies in the North. The relief feature of the country comprise of low lying areas between 800 to 900m above sea level in the South, and the low lying areas of the Kariba basin to the North, the high plateau with more than 1 000m above sea level stretches from South East to the North. The Eastern Highlands has some areas lying at 1 200 m above sea level. High rainfall is commonly received in the Eastern Highlands and North areas in the North East and also along the central high plateau. Rain season is experienced between November and April; the cold season is between May and July. Hot dry season begins in August and stretches to October.

Zimbabwe has a population of 11 634 663 people with a distribution of 5 631 426 males and 6 003 237 females according to the national census of 2002. The majority (80%) of these people live in rural areas. The average size of households is 4.4 according to Zimbabwe National census summary report 2002. The Government of Zimbabwe is struggling to ensure equity in the distribution of natural resources among all races that inhabits the country. These efforts include the land reform process that is underway in the country to ensure equal distribution of fertile lands among the people of Zimbabwe. The government is also fighting repeated droughts by constructing dams for irrigation purposes. HIV, TB and malaria are the main big three diseases that affect the country. The government of Zimbabwe has considered

* National Institute of Health Research, Ministry of Health, P.O. Box CY 573, Causeway, Harare, Zimbabwe; nicholasmidzi@yahoo.com

control of these big three disease as matters of high priority in its national agenda. Research is therefore prioritized on these three enemies of health.

Biomedical research in Zimbabwe includes national research programmes on which research is conducted at national level mainly focused on communicable diseases HIV, TB, Malaria, diarrhea and dysentery, neglected diseases (schistosomiasis and geohelminthiasis) and non-communicable diseases, reproductive health and nutrition (Mduluza et al. 2001; Sibanda et al. 2003). There is also private institutional research that also focuses on main diseases outlined above.

The Study Population and Communities Available for Biomedical Research

The population available for research includes both urban and rural community people. Both adults and young children (mainly school aged children) are very co-operative and actively participant in any biomedical research if awareness communication is provided through their local leaders and other community representatives (Machekano et al. 2002; Nilses et al. 2000).

Fears and Beliefs

Fears are commonly encountered when invasive research that include, collection of blood and samples/specimens from genital organs and if research is conducted on females especially in married women who may need permission to participate in such biomedical research from their spouses.

In any community, research that is conducted by foreigners without the involvement of known local people is likely to frighten participants. It is therefore advised that foreign researchers incorporate local people of the community they intend to conduct research on, such as Environmental Health Officers (EHOs), Environmental Health Technician (EHTs), Community nurses, health workers or teachers. Normally the communities know these people and respect them since in all cases of epidemic diseases outbreaks such cadres are routinely met by the community during intervention. They can also speak local language well so they can communicate well the proposed study in local language. International investigators that may need to conduct research in the country also need to engage local scientists, from a recognized institution.

Religion in Zimbabwe plays a part in research as well, some people believe in traditional way of life, appeasing their ancestors for their survival or to live well. Others are Christians whilst others are Moslems. But many are either traditional people or Christians. Some Christians like the apostolic sects in some areas do not allow their members to get convectional medicine. It is therefore difficulty for investigators to conduct research on these people in a similar way it could be difficulty to have these members of the society to get treatment against epidemic diseases. They generally do not agree to participate in any study and give out their samples for any laboratory analysis (Mduluza et al. 2001; Mutapi et al. 2005).

Economic and Social Status

Zimbabwe is an agro-based country whose people earn their living through farming that is divided into subsistence and commercial farming. Crops commonly grown in the country at large scale include maize, tobacco, wheat, soya beans, ground nuts, rapoko and millet. There are sugar plantations in the low veldt (areas in Chiredzi). Banana and tea plantations are found in the eastern highlands which receives constant rainfall almost through out the year. Tobacco is also grown mainly on the high veldt and eastern highlands. Mining, tourism and other processing industries are other major sources of income for people.

Political Status

Zimbabwe is a very stable country when it comes to political status. The only wars that were waged in the country were the liberation war, (The Second Chimurenga) intensively fought from 1963 to 1979 when ceasefire and election in 1980 led to the majority rule. The Second Chimurenga, followed the first Chimurenga that constitute a series of early resistance by Zimbabweans to foreign occupation and rule. Following the second Chimurenga, there has not been any armed war in Zimbabwe. Parliamentary and presidential elections are conducted peacefully at the end of each term as has been witnessed in successive terms since 1980 until now. The political structures in relation to the community begin at the Minister of Local Government, Governors, District Administrators, Councilors, and Ward coordinators. The traditional structures begin with the Chiefs (Madzishe in local language) and goes to village head men (maSabhuku in local language). Political leaders work hand in hand with traditional leaders at each level. It is very important when researchers enter the communities meet these political and traditional leaders, explain their research objectives and seek permission to work with the community. The political leaders would want assurance that research proposed to be conducted in their communities does not end up interfering with the political set up of the community. If all is fine with investigators, these cadres are very co-operative and are always ready to assist the entry and introduction of researchers in their community structures.

The Community Hierarchy for Biomedical Research

Following obtaining of the approval/clearance letter to conduct research in Zimbabwean population from the Medical Research Council of Zimbabwe (MRCZ), researchers have to go through several stages until they finally get to the prospective research candidates. It is often encouraged that the foreign researchers team up with local (Zimbabwean) researchers in order to instill confidence in the study population and also to expedite approval processes at every stage of entry into the community. However, involving local researchers will not substitute the requirements of approval procedures as one goes into the community. It simply minimizes the delays that may be caused at each stage in the community structures.

Participation of Adults

There are two scenarios used to access the prospective subjects depending on age of the study population specified in the research protocol. Planning to conduct research on adults in the community, involves getting permission from the MRCZ to conduct such research following the review of the research protocol. From this stage, researchers are expected to notify the Local Government through the Governor either in writing or by visiting the offices and discussing the project. The Governor reserves the right to interview the investigators. Investigators are expected to produce a clearance letter from the MRCZ that indicate permission granted to conduct such research in the specified community. The copy letter is filed at the Governor's office that in turn ensures that the local community leaders provide assistance to researchers as they get into the community provided the governor is convinced that the research will not bring problems to the local people. A letter may be obtained from the Governor to the District Administrator (DA) seeking permission and also to facilitate entry of the researchers into the district.

The DA reserves the right to know the objectives of the study and their relevance in the area of study too. This becomes simple when the research has a simplified consent form and summary of the proposal. The DA may write letters to the local community leaders who include the chiefs and or ward councilors to facilitate entry of and co-operate with researchers as they get into the community. The councilors will then inform the village headmen and the community of the intended research project. At this stage the researchers should have linked up with familiar community representatives such as Environmental Health Technician (EHTs) or Environmental Health Officers (EHOs), Community nurses or community health workers. Normally the community is familiar with these people and do respect them since in all cases of epidemic disease outbreaks such cadres are met by the community when they provide first line intervention or management of diseases (such as malaria cases). The presence of these cadres at meetings for focus group discussions (FGDs) when the project is being introduced to the community will demystify fears of the community that may be associated with entry of unfamiliar or foreign workers in their structures. It is particularly important to arrange for such meetings with the community as the project is introduced into the community. Creation of such meetings facilitates focus group discussion, the platform that may be taken to demystify fears from the community and to clearly explain the objectives of the study to participants, although individual visits may be done later according to the design of the protocol and the confidential concerns that may need to be ensured.

At the village level the researchers will approach the village headmen who will also allow the investigators into the village following notification. At the household level, the researchers have to seek permission to conduct research on specified individuals in the family or any family head in the case of single parents or guardians in the case of minors. Following this, informed consent has to be sought from the specific research candidate who will be referred to as research participants. The prospective research participants have to be fully informed about the objectives of the study, the procedures to undergo during research activities, any foreseeable risks, benefits associated with research and the community, any treatment available in the event that adverse reactions or injury do occur due to research. All the informed consent procedures have to be conducted in simplified terms and in local language to ensure that the participant makes informed decision to participate without undue influence, intimidation or inducement. The voluntary nature of the project should be explained to each individual. Before the research is conducted each individual selesmall tumor size, high proportion of well/moderately differentiated tumors and consequently a lower NPI value. In contrast, patients with luminal B tumors were more likely associated cted for the study is expected to sign consent form to indicate competence in understanding the objectives of the project and is willing to participate.

On the other hand, since any kind of research on human subject has something to do with their health, before research is conducted on individuals in the community, permission is also sought to conduct research from the Provincial Medical Directorate (PMDs) through to the District Medical Officer (DMO), who also reserve the right to interview the researchers and also to review the proposal. This is very important since occurrence of any adverse reactions during research would need immediate local medical attention and explanation. The local health authorities would also want to know the implication of research on the health of the community and may provide useful advice to researchers. In most situations from PMD and DMO the researchers team up with the local community health workers especially the environmental health technicians or village health workers, the cadres may be present when the researchers seek community permission to participate in research so that they can help explain to the community the nature of research using local language. The community, as outlined above trusts these people and may be familiar of the link between the local health issues and the research theme. The procedure of seeking permission from health authorities should be done apparel to that procedure taken to seek permission through the hierarchy of the Local Government in order to avoid delays as in all cases these offices are always located in the same premises (at provincial or district level).

Participation of School Children

If research participants are school children, then in addition to the two pronged procedures of obtaining permission to enter into the community as explained above, after obtaining clearance from MRCZ, the researchers are expected to seek permission to conduct research in schools from the Provincial Education Directorate who may also want to know the nature of the project, duration of the project and frequency of the follow up surveys at the schools. Education directorate may wish to agree with the researchers on the time expected working at each school in order to avoid disruption of school activities. The researchers are also expected to seek permission from the District Education Officer (DEO), who usually notifies school heads, seeking their assistance or introducing the research team to the school. Upon arrival at the school, the researchers have to seek permission from the school head to conduct their research in the school. Following discussions with the school head in the presence of local health cadres, permission will be sought from the parents and guardians of the children, who may already have been sensitized by the village health workers. The children aged between 7-12 years, are expected to complete the assent form whilst the parents complete the consent forms stating that they permit their children to participate in research after understanding the nature of the research to be undertaken on their children. The assent form must indicate the voluntary nature of the research, that is to say the child is free to withdraw from the study at any time he or she feels like (Mduluza et al. 2001; Mutapi et al. 2000; Ndhlovu et al. 1996). The assent form should be simplified enough to ensure that children comprehend the nature of the research and the benefits associated with conducting such research on them and the society.

National Bioethics Process in Evaluating Biomedical Research into the Country

The Medical Research Council of Zimbabwe (MRCZ) is the only ethical committee or National top review board accredited in Zimbabwe. It consists of a 14-member council who represent various stakeholders in accordance with the piece of legislation, which constitutes the MRCZ. Members are nominated by various constituencies and are then appointed by the Minister of Health and Child Welfare for a four year term. Stakeholders represented by the MRCZ Council include Ministry of Health and Child Welfare, Universities, National Institute of Health Research, Medicines Control Authority of Zimbabwe, Zimbabwe Medical Association and the Permanent Secretary, Ministry of Health and Child Welfare.

The MRCZ regulates the conduct of biomedical research in Zimbabwe. Trials that involve drugs and biologicals, the Medicine Control Authority of Zimbabwe (MCAZ) regulates medical devices, which is a statutory body responsible for the registration of drugs, biologicals and medical devices. For such trials investigators may only proceed with the trial once they obtain approval from both the MRCZ and the MCAZ. The MRCZ, which also serves as the National Ethics Committee (NEC) was established in 1974 in terms of the Research Act of 1959 and Government Notice Number 225 of 1974 in order to provide health researchers and institutions in which health research is conducted, with independent ethical advice on research conducted by those researchers by/within those institutions. The MRCZ was established and supported by the Government of Zimbabwe through the Ministry of Health and Child Welfare. It is composed of scientist, medical experts, patient representatives and the community representatives. It is independent in its reflection and advice.

Institutional Review Boards (IRBs)

Institutions that include National Institute of Health Research, Biomedical Research and Training Institute, Universities and Hospitals where biomedical research can be conducted do have their own institutional review boards consisting of members selected to review research protocols and to give comments before the protocols are finally submitted to the MRCZ. The IRBs do not have the powers to approve the protocol for conduct. Final approval of any research protocol is granted by MRCZ although the IRBs can review them at institutional level. Following approval of the scientific or health research protocol by IRBs, the protocol is submitted to MRCZ for final review and approval. In other institutions the review of proposals is conducted by seasoned researchers that are selected into the Research and Development committees. These Research and Development committees perform their duties in the similar way as IRBs.

Documents for Submission of Proposal to MRCZ

The following documents make up a complete application package that should be submitted to the MRCZ office; normally enough copies for distribution to all involved parties in evaluation of the application:

- Twelve copies of application forms duly completed.
- Twelve forms of protocol summary less than three pages in length.
- Four copies of complete research protocol in general funding format.
- Four copies of consent forms in English and local language of the study population.
- Four copies of drug brochure or any complementary information if applicable.
- Four copies of questionnaires being administered during the study.

Other Requirements

- A registration fee which is payable on submission of research proposal for review.
- The proposal by foreign-based researchers has to be submitted through the Research Council of Zimbabwe including a registration fee payable to the research council of Zimbabwe.
- Documentary evidence of local institutional support and clearance will help to expedite approval of proposal submitted by foreign researchers.
- For research involving the testing of drugs and medical devices, permission also has to be obtained from the Medicines Control Authority of Zimbabwe.

MRCZ Decision on Submitted Protocols

Applications are acknowledged and the researchers are informed of the review date. The protocol will be sent to at least four specialized reviewers who will be asked to give comments leading to approval or disapproval of the research conduct. The researchers will also be communicated regards the incompleteness of the application. The outcome of review will be communicated within a week after the MRCZ meeting.

Researchers May Be Asked to Respond to Issues Raised in the Meeting if Required

If the research protocol is approved a clearance letter will be written to the project investigator. This letter will serve as a proof of permission granted to the researcher (s) to conduct biomedical research in the specified site/area. However this document will not substitute the need for permission to get into the community nor does it substitute the need for the researcher to seek informed consent from the research participants.

Guidelines on the Conduct of Research in Zimbabwe

The requirements from any investigator willing to conduct research in Zimbabwe have to be certified by the Medical Research Council of Zimbabwe. Any research projects involving human subjects, whether as individuals, communities including the use of foetal material, embryos and tissues, from the recently dead, supported or undertaken by any institution, researchers wherever conducted in Zimbabwe, shall be reviewed by the Ethical Review Committee at the institution at which it is going to be conducted as well as the Medical Research Council of Zimbabwe before the study begins. The basic principles of the ethical review committees are to:

- Determine that all proposed interventions, particularly administration of drugs and vaccines or the use of medical devices or procedures under development are acceptably safe to be undertaken on humans, determine that the proposed research is scientifically sound or to determine that another competent expert has done so.
- Ensure that all other ethical concerns arising from the protocol are satisfactorily resolved both in principle and in practice.
- Consider the qualifications of the investigators, including education in the principles of research practice and the conditions of the research site in the context of safe conduct of the trial.

The informed consent of all humans subject participating in the research must be sought before the research activities are initiated. Informed consent is a decision to participate in research taken by a competent individual who have received the necessary information, adequately understood the information and who after considering the information has arrived at a decision without undue influence or inducement or intimidation.

Waiver of the Consent Requirement

Investigators must never conduct research without obtaining informed consent from each subject unless they receive explicit approval from the ethical review committee as well as the MRCZ.

Reviewing Consent

When material changes occur in the condition or the procedures of the study, the investigator should once again seek informed consent from the subjects.

Cultural Considerations

In some cultures the investigator may enter the community to conduct research or approach the prospective subjects for their individual consent only after obtaining consent from the community leader, a council of elders or other designated authority. Such customs must be respected. In no case however may the permission from the community leader or other authority substitute for an individual's provision of an informed consent.

Use of Medical Records and Biological Specimens

Medical records and biological specimens taken in the course of clinical care may be used for research without the consent of the participant only if an ethical review committee and MRCZ has determined that the research poses minimum risk, that the rights or interest of individual patients will not be violated, that their privacy or confidentiality are assured.

Secondary Use of Medical Records and Biological Specimens

Investigators may want to use the records or specimen used by other researchers or collected for use, in another institution or country. This raises issue of whether the specimens contains the personal identifiers, or can be linked to such identifiers. If original informed consent was required to authorize the original collection of the specimen for research purposes. Secondary uses are generally constrained in the conditions specified in the original consent. Such specimens can only be used for further analysis after obtaining permission from the ethics committee and the MRCZ.

Essentials of Informed Consent in Zimbabwe

Comprehension

Investigator must ensure that the informed consent is clearly comprehended by the subject or guardian. Purpose of research must be clearly explained (Glossary). Describe the procedure in simple terms that the subjects would be expected to undergo. Identify any procedures that are experimental. Indicate type and frequency of monitoring during and after the study. Length of time the subject is expected to participate should be explained. Benefits of the study must be shared with the participants, and society. Explanation should be given for all foreseeable risks to the subjects. Treatment available in the event of adverse reaction or injury as a result of being a participant in the study needs to be explained.

Guidelines for Researchers Conducting Research with Children in Zimbabwe

In general all children who are asked to participate in biomedical research are expected to give their assent. Examples of assent, consent attached at end in the glossary.

The following definitions need to be kept in mind:

(a) Assent means the child's affirmative agreement to participate in research. Mere failure to object should not be misconstrued as assent.

(b) Therapeutic research refers to research that holds out the prospect of direct benefit to the individual subject that is only available through participation in the research study.

(c) Non-therapeutic research refers to the research that offers no benefit or only indirect benefit to the individual subject.

Monitoring of Biomedical Research

Monitoring of biomedical research in Zimbabwe is done through three ways. These include field visits by MRCZ officials, request for progress reports and use of measures to guard against deviations from protocols. Field visits are carried out by MRCZ officials. These visits are made to monitor biomedical research process practically. Usually the investigators are not warned about the visits. This is done to ensure that the conduct of research is done according to the protocol as indicated in the research proposal. The MRCZ officials may also ask participants on the procedures they are being subjected to. Reports from such field visits will be used to assess biomedical research and any deviations in methodologies observed with no prior notification and approval from MRCZ, will not only lead to arrest of investigators but also to cancellation of license to conduct research in Zimbabwe.

Progress reports are requested quarterly from investigators. The reports should indicate any problems encountered during the period of the biomedical research conducted. If there are practical methods to be changed or if there is some need to change the methodologies due to practical experiences then this should be highlighted with justification in the progress report. Ultimately this means any changes sought can only be made following review of the progress report by the MRCZ council. The MRCZ currently request investigators to renew their approvals annually if the project is undertaken for more than a year.

Current and Future Status of Biomedical Research in Zimbabwe

When dealing with school aged children normally the investigators visits children in schools to enable accessing as many participants as possible. Incentives that are normally given to children include sport uniforms and contributing towards development of the school. On the other hand any results coming out of the study should be used for community intervention measures, thus benefiting the society. Basically there is no set kind of incentives as the ethical review boards discourage any form of inducement of the community into the study. Treatment of some minor illness and the diseases under investigation is given to the study participants. Sometimes, studies that have to do with schistosomiasis control using school based approach, it is mandatory that during treatment days children are fed before they get treatment. In this case although children are urged to bring food and eat some food at home, researchers ensure that food is provided on the day of treatment in case some children may not have adequate food with them or may not have taken any food at home prior to treatment. Bread and orange crush juice are normally given to children together with treatment.

The Adult Population Participating in the Study

If the adult participants are required to visit a station or a clinic for biomedical procedures then they would be given transport refund and also snacks/food during waiting time. This helps mobilize the community if the study is a follow up design. This applies to the adult population in urban settlements who may have to travel some distance to the site for biomedical procedures. There is no stated amount of payment for waiting time to participants who spend their time on the day of follow up when they have to leave their routine duties to attend to the session for biomedical procedures. Though inducement is not permitted in biomedical research, some kind of compensation should be given to participants for their lost time and to boost their participation moral during follow-up. However because the local research scientist rely heavily on donor funds for research, a balance has to be made that ensures that the budget for incentives does not end up taking a sizeable percentage of the total budget of research. Often donors are willing to spend their money on core research rather than overheads and incentives.

Given that the standards of incentives in the United State require that one can even publish the amount paid to participants in monetary value, if the same is done in developing countries then there is fear of competition for participants by researchers when incentives are liberalized with no limit in the third world. A condition will arise when local researchers will end up accessing the study population through the projects linked to investigators from rich funders. There could therefore limit local researchers who often struggle to get minimal funding to conduct biomedical research from local funding agencies or donors. All these are issues to be considered in the context of sustaining local biomedical investigators and students. However detailed analysis of the concerns to compensation to partipants is necessary since the practice may not continue for long without the participants from developing contries realizing that the incentives/compensation is not similar in developed and developing countries.

References

New rules for biomedical research projects. *Nature 246:* 239-41, 1973

International ethical guidelines for biomedical research involving human subjects. *Bull. Med. Ethics:* 17-23, 2002

Bandewar, S.: Cultural barriers, 'competence' and informed consent in population-based surveys. *Issues Med. Ethics. 11:* 49-51, 2003

Barker, J. H. and Polcrack, L.: Respect for persons, informed consent and the assessment of infectious disease risks in xenotransplantation. *Med. Health Care Philos. 4:* 53-70, 2001

Corrigan, O.: Empty cthics: the problem with informed consent. *Sociol. Health Illn. 25:* 768-92, 2003

Creed-Kanashiro, H., Ore, B., Scurrah, M., Gil, A., and Penny, M.: Conducting research in developing countries: experiences of the informed consent process from community studies in Peru. *J. Nutr. 135:* 925-8, 2005

Crigger, N. J., Holcomb, L., and Weiss, J.: Fundamentalism, multiculturalism and problems of conducting research with populations in developing nations. *Nurs. Ethics. 8:* 459-68, 2001

Deschenes, M., Cardinal, G., Knoppers, B. M., and Glass, K. C.: Human genetic research, DNA banking and consent: a question of 'form'? *Clin. Genet. 59:* 221-39, 2001

Foster, M. W., Bernsten, D., and Carter, T. H.: A model agreement for genetic research in socially identifiable populations. *Am. J. Hum. Genet 63:* 696-702, 1998

Helgesson, G. and Johnsson, L.: The right to withdraw consent to research on biobank samples. *Med. Health Care Philos. 8:* 315-21, 2005

Jamison, J. R.: Informed consent: an Australian case study. *J. Manipulative Physiol. Ther. 21:* 348-55, 1998

Jiwa, M.: The recruitment of non-English speaking subjects into human research. *J. Med. Ethics 25:* 420-1, 1999

Jokinen, P., Lappalainen, M., Merilainen, P., and Pelkonen, M.: Ethical issues in ethnographic nursing research with children and elderly people. *Scand. J. Caring Sci. 16:* 165-70, 2002

Katz, J.: The Nuremberg Code and the Nuremberg Trial. A reappraisal. *Jama 276:* 1662-6, 1996

Kishore, R. R.: Biomedical research and mining of the poor: the need for their exclusion. *Sci. Eng. Ethics. 12:* 175-83, 2006

Kopelman, L. M.: Minimal risk as an international ethical standard in research. *J. Med. Philos. 29:* 351-78, 2004

Lavori, P. W., Krause-Steinrauf, H., Brophy, M., Buxbaum, J., Cockroft, J., Cox, D. R., Fiore, L., Greely, H. T., Greenberg, H., Holmes, E. W., Nelson, L. M., and Sugarman, J.: Principles, organization, and operation of a DNA bank for clinical trials: a Department of Veterans Affairs cooperative study. *Control. Clin. Trials. 23:* 222-39, 2002

Machekano, R., Bassett, M., McFarland, W., and Katzenstein, D.: Clinical signs and symptoms in the assessment of immunodeficiency in men with subtype C HIV infection in Harare, Zimbabwe. *HIV Clin .Trials. 3:* 148-54, 2002

Macklin, R.: Understanding informed consent. *Acta Oncol. 38:* 83-7, 1999

Macpherson, C. C.: To strengthen consensus, consult the stakeholders. *Bioethics. 18:* 283-92, 2004

Marshall, E.: Research ethics. NIH examines standards for consent. *Science 280:* 1688, 1998

Marshall, P. A.: Human rights, cultural pluralism, and international health research. *Theor. Med. Bioeth. 26:* 529-57, 2005

Mduluza, T., Ndhlovu, P. D., Madziwa, T. M., Midzi, N., Zinyama, R., Turner, C. M., Chandiwana, S. K., Nyazema, N., and Hagan, P.: The impact of repeated treatment with praziquantel of schistosomiasis in children under six years of age living in an endemic area for Schistosoma haematobium infection. *Mem. Inst. Oswaldo Cruz. 96 Suppl:* 157-64, 2001

Molyneux, C. S., Peshu, N., and Marsh, K.: Trust and informed consent: insights from community members on the Kenyan coast. *Soc. Sci. Med. 61:* 1463-73, 2005

Morens, D. M.: Should the Declaration of Helsinki be revised? *N. Engl. J. Med. 341:* 1851; author reply 1853, 1999

Mutapi, F., Burchmore, R., Mduluza, T., Foucher, A., Harcus, Y., Nicoll, G., Midzi, N., Turner, C. M., and Maizels, R. M.: Praziquantel treatment of individuals exposed to Schistosoma haematobium enhances serological recognition of defined parasite antigens. *J. Infect. Dis. 192:* 1108-18, 2005

Mutapi, F., Ndhlovu, P. D., Hagan, P., and Woolhouse, M. E.: Anti-schistosome antibody responses in children coinfected with malaria. *Parasite Immunol. 22:* 207-9, 2000

Ndhlovu, P., Cadman, H., Vennervald, B. J., Christensen, N. O., Chidimu, M., and Chandiwana, S. K.: Age-related antibody profiles in Schistosoma haematobium infections in a rural community in Zimbabwe. *Parasite Immunol. 18:* 181-91, 1996

Nilses, C., Lindmark, G., Munjanja, S., and Nystrom, L.: A community based study of HIV in women in rural Gutu District, Zimbabwe 1992 to 1993. *Cent. Afr. J. Med. 46:* 32-7, 2000

Pang, T.: Setting global health research priorities: ethics should also guide global health research. *Bmj 326:* 1399, 2003

Patil, R.: Research in a tribal community. *Indian J. Med. Ethics. 1:* 122-3, 2004

Peppin, P.: Power and disadvantage in medical relationships. *Tex. J. Women Law 3:* 221-63, 1994

Phanuphak, P.: Trials in developing countries. *Asian Pac. J. Allergy Immunol. 16:* 137-9, 1998

Phipps, E., Fleetwood, J., and Piraino, A.: Research ethics and clinical trials: fostering collaboration with research participants. *Account Res. 6:* 215-26, 1999

Phoenix, J. A.: Ethical considerations of research involving minorities, the poorly educated and/or low-income populations. *Neurotoxicol. Teratol. 24:* 475-6, 2002

Pickering, N.: Indigenous peoples and medical research. *N Z Bioeth J 4:* 2, 2003

Pickworth, E.: Should local research ethics committees monitor research they have approved? *J. Med. Ethics. 26:* 330-3, 2000

Pickworth, E. and Brazier, M.: Fees and research ethics committees. *Bull. Med. Ethics No. 151:* 18-20, 1999

Pinching, A. J.: Live attenuated vaccine trials in medically informed volunteers: a special case? *J. Med. Ethics. 26:* 44-6, 2000

Pinkerton, C. R., Ablett, S., Boos, J., and Philip, T.: Ethical approval for multicentre clinical trials in children. Contrasting systems in three European countries. *Eur. J. Cancer 38:* 1051-8, 2002

Pletscher, A.: Research and society: a relationship with two faces. *J. Recept Res. 13:* 1-18, 1993

Pope John, P., II: The profit motive behind too much medical research. *Origins. 31:* 754-5, 2002

Pope John, P., II: Biomedical research: benefits, limits of science. *Origins. 32:* 673, 675-6, 2003

Resnik, D. B., Zeldin, D. C., and Sharp, R. R.: Research on environmental health interventions: ethical problems and solutions. *Account Res. 12:* 69-101, 2005

Rhodes, R.: Rethinking research ethics. *Am. J. Bioeth. 5:* 7-28, 2005

Saha, S.: Challenges in research in tribal communities. *Indian J. Med. Ethics. 1:* 125, 2004

Salinas Mulder, S., Rance, S., Serrate Suarez, M., and Castro Condori, M.: Unethical ethics? Reflections on intercultural research practices. *Reprod. Health Matters. 8:* 104-12, 2000

Sibanda, E. N., Stanczuk, G., and Kasolo, F.: HIV/AIDS in Central Africa: pathogenesis, immunological and medical issues. *Int. Arch. Allergy Immunol. 132:* 183-95, 2003

Sleeboom, M.: The Harvard case of Xu Xiping: exploitation of the people, scientific advance, or genetic theft? *New Genet. Soc. 24:* 57-78, 2005

Sobel, M. E.: New challenges for biomedical research utilizing human biological materials. *J. Mol. Diagn. 2:* 1-4, 2000

Srinivasan, S.: Research on public health interventions in poor countries. *Issues Med. Ethics. 9:* 118-9, 2001

Steinberg, K. K.: Ethical challenges at the beginning of the millennium. *Stat. Med. 20:* 1415-9, 2001

Turner, L.: From the local to the global: bioethics and the concept of culture. *J. Med. Philos. 30:* 305-20, 2005

Wicher, C. P. and Michalek, A. M.: When is informed consent not enough? *J. Cancer Educ. 20:* 9-10, 2005

Zion, D.: Community without communitarianism: HIV/AIDS research, prevention and treatment in Australia and the developing world. *Monash. Bioeth. Rev. 24:* 20-31, 2005

Zion, D., Gillam, L., and Loff, B.: The Declaration of Helsinki, CIOMS and the ethics of research on vulnerable populations. *Nat. Med. 6:* 615-7, 2000

In: A Gateway to Biomedical Research in Africa
Editor: Takafira Mduluza, pp. 131-139
ISBN: 1-60021-444-4

Chapter XIII

The Republic of Ghana, West Africa: Steps in Doing Health Research in Ghana

Collins Stephen Ahorlu[*]

Brief Introduction on Country Biomedical Research Status

Ghana is a developing country located in West Africa. It has a population of about 20 million people. The majority (over 60%) of the people live in rural areas surviving on agro-based economy, mainly subsistence farming, which is labour intensive. The government of Ghana is doing its best to achieve the millennium development goals. However, diseases, poverty and lack of education continue to work against the developmental efforts of the state. To combat diseases afflicting the people of Ghana requires research, both basic and applied.

The research agenda is divided into two broad areas – Basic Health Research and Applied Health Research. However, the MoH as an executive arm of Government health policy formulation concentrates more on applied health research leaving basic research to other Institutions like the Universities and private health related industries, especially in the area of product development.

[*] Research Fellow, Noguchi Memorial Institute for Medical Research, University of Ghana P. O. Box LG581, Legon, Accra, Ghana, West Africa. Tel: +233-21-501178 (office); +233 208195705 (mobile); +233-21-502182 (fax); Email: cahorlu@noguchi.mimcom.net or Collins.ahorlu@unibas.ch Research Fellow, Noguchi Memorial Institute for Medical Research, University of Ghana P. O. Box LG581, Legon, Accra, Ghana, West Africa. Tel: +233-21-501178 (office); +233 208195705 (mobile); +233-21-502182 (fax); Email: cahorlu@noguchi.mimcom.net or Collins.ahorlu@unibas.ch

Applied Health Research

Currently, six operational health research priority areas are identified in the national policy as follows:

1. Access to Health Services: Research in this area seeks to address the barriers to uptake of health care services. Some of the barriers identified include, geographical or physical barriers – location of facilities becomes very important; cost of service provision – user fees; cost of transportation and travel time to facilities; sociocultural barriers – opening hours of clinics and people regarding some conditions as not for hospital treatment. This access problem would need to deal with equity issues, identifying vulnerable groups among others.
2. Quality of Health services: This requires identifying factors affecting the quality of service being provided to the public. Clients' perceptions and expectations as well as provider attitude are critical in this type of research.

    a) Efficiency in the use of Health Resources: The appropriateness of service delivery and its ability to impact on morbidity and mortality is considered here. The allocation of resources, human and materials, as well as developing integrated approach to service delivery is also considered here.
    b) Linkages in the Health Sector: This research agenda focuses on establishing linkages among various service providers, thus finding ways to integrate the work of various providers, including traditional providers, to better serve the interest of the population.
    c) Health Financing and Health technology: The sustainability of every health care delivery depends on how it is financed. Advances in medical technology come with cost implications. This agenda deals with sources of financing health services and financing options available including the payments of user fees. Identifying cost - effectiveness of service delivery and considers the options of promotion, preventive health delivery as an alternative to curative care. Determining the cost-effectiveness of medical technologies through field trials. The economic use of service which goes beyond the direct payments will also be studied under this priority.
    d) Selected Health programmes, Morbidity and Mortality studies: Under this priority agenda, Epidemiological studies of infectious diseases and other health problems of public health importance are carried out.

Basic Health Research

The agenda for this component of the national health research policy concentrates on the diseases of public health importance in Ghana. This policy direction concerns all aspects of health research. Some of the areas of emphasis include: Bacteriology, Clinical Research, Clinical Pathology, Electron Microscopy, Entomology, Epidemiology, Parasitology,

Pharmacology, Immunology, Virology, Haematology, Molecular research, and all aspects of social science. These fields are applied to research into many conditions of public health interest and priority, some of which are listed below:

Malaria, Tuberculosis, ARI, Diarrhoea, Sexually transmitted infections, HIV/AIDS, Leprosy, Guinea worm infections, Onchocerciasis, Schistosomiasis, Buruli ulcer, EPI services, Reproductive Health, Maternal and child health, Prevention and Control of Micronutrient Deficiencies, etc.

These Health Research Priorities are used to guide and inform the development of research portfolios and research investments for Ghana.

Population and Communities Available for Biomedical Studies

Ghanaians are very friendly people, always cooperative and ready to participate in any beneficial biomedical research. However, there are steps to follow when entering inton any Ghanaian community. The first step involves meeting with community leaders, mostly chiefs and elders, to discuss the propose study with them to seek for their cooperation and acceptance. The second step involves meeting the entire community at a durbar to be organized by the chiefs and elders to discuss the proposed study with the community at large to seek for their cooperation and participations. At these meetings questions about the project are asked and clarifications sought. When this is done, then, one can be sure of the support, participation and success of the project since it has the backing of the people. These meetings must be preceded by meetings with the District Health and political officials and in some cases the regional officials to seek for professional and political support for the project.

However, studies involving blood collection have to be well explained to participants to allay their fears due to some local beliefs about blood and issues having to do with HIV/AIDS testing. Ghanaians are generally receptive to foreigners but it is always good to have a Ghanaian in the research team because of language barriers and cultural sensitivities.

Political Stability in Ghana

Ghana attained independence in 1957 and gained a republic status in 1960 under the leadership of the late President Osagyefo Dr. Kwame Nkrumah of blessed memory. Ghana after some turbulence periods has been very stable in the past 25 years and is rightly considered one of the growing democracies in sub-Saharan Africa, practicing multi-party democracy. Ghana is sometimes acclaimed as the "Island of peace" in the midst of civil strives in the West African sub-region. However, the country is yet to convert this political stability into improve health status of the populace, food security for the people and quality education, especially in the public educational institutions.

National Bioethics Processes in Evaluating Biomedical Research in Ghana

Scientific and Technical Committee (STC)

A proposal is first submitted to the Secretariat of the STC. This committee looks at the scientific merits of the research proposal and satisfies itself of the application of the scientific rigor in terms of sound methodological design and the relevance of the toll for data collection. It also looks at the originality and its contribution to scientific and applied knowledge. This committee exists in most health research institutions in the county and could be utilized by any person wishing to conduct health research in Ghana. After submission the proposal is distributed to committee members who read and prepare their comments that are discussed at the committee's meetings.

Institutional Review Boards (IRB)

The IRB is the top most review board which includes members from various institutions in the country, some of which are, Ministry of Health/Ghana Health Service, Health research institutions, Council for Scientific and Industrial Research, Faculty of law, Ghana Medical School, Social Sciences and Religious Bodies. The IRB meets once every two months provided materials have been submitted for review. IRBs are also available in most health research institutions including the MoH and could be used by anybody wishing to conduct health research in Ghana.

New Protocol Submission

The principal investigator is responsible for submitting the following material for a review to take place:

i. Summary of the Research Protocol
ii. A full protocol previewed by a scientific committee with the comments and giving details of methodology, materials and sample collection.
iii. Enrolment forms
iv. Questionnaires
v. Consent forms
vi. Curriculum vitae of investigators

Process of Submission

a. Before submission for ethical review, the investigator should have submitted a protocol to the Scientific and Technical Committee (STC) of the Institute where the work will be based for clearance. The investigator should certify that the proposal has gone through STC review by signing a declaration on the cover page.

b. New protocols should be submitted to the chairman of the IRB through the IRB administrator not later than 2 weeks prior to an IRB meeting (new proposals should be submitted with a cover letter from the investigator)

c. Upon receipt of the protocol, the IRB administrator will assign a code number to the protocol. The Administrator acknowledges receipt of the protocol.

d. Failure to submit all required materials may result in delay in consideration of the protocol.

Monitoring of Biomedical Research

Monitoring and evaluation is done in three ways as follows:

- Research Process
- Monitoring of Ethical issues
- Evaluation of Outcomes

Research Process Review

Here the research team and the IRB monitors take part in the process review to be sure whether the research is carried out as proposed. It also involves asking questions such as, will the study ends within the proposed time frame? Does the study need to change some of the research methods midway through data collection? If so what are the implications on the review processes? Are the reasons for a change clearly documented and justified? Would the change require a new review processes?

Ethics and Human Rights

This is the most important aspect in biomedical research monitoring. Researchers will be monitored to see whether they follow the ethical procedures described in the proposal. All the Ethical commitments spelt out in the proposal must be adhered to. Any violation could lead to the termination of study. In the extreme, the research team could be sued at law court for ethical and human rights violations. To achieve this, the research team represented by the principal investigator (PI) must submit periodic reports to the IRB for review to be sure that the research team is following the laid down ethical procedures as enshrined in the proposal as well as meeting the IRB's ethical standards.

Outcomes Evaluation

It is important to evaluate the success or failure of a research against what it sets out to achieve at the end of the study. This is done through the submission of the final report to STC and IRB as part of the documentation and dissemination of results/findings. Findings must also be disseminated to participating communities.

The Hierarchy for Obtaining Permission

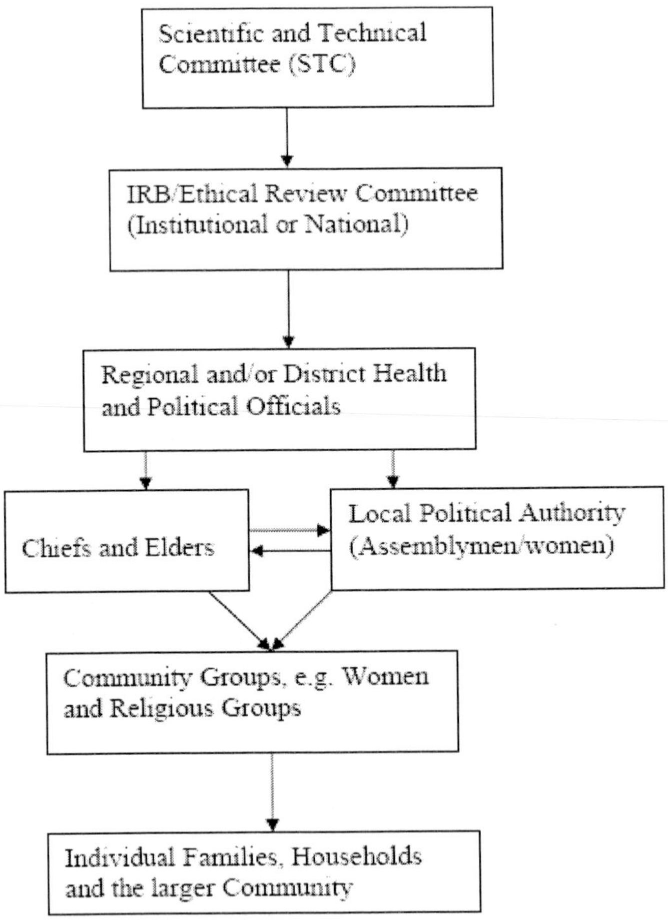

Generalized summary diagram on the hierarchy for obtaining permission.

After receiving STC approval and IRB Ethical Clearance Certificate the investigators need to ask permission to get into the district. In most cases, permission is sought directly from the District Health Management Team (DHMT) which in turn informs the regional health directorate. In some cases however, permission has to be sought from the regional health directorate before entering into the district. It must be emphasized that all these depend on how much consultation was done or collaborations that has been established at the initial stage of the protocol development.

The next institutions or persons to contact for discussions are the chiefs and elders and or the assembly man or woman representing the community at the District assembly, which is the highest political authority in the district. Meeting these two groups is very essential for various reasons but most importantly for community mobilization for the study. Once the study receives the blessings of the chiefs and elders as well as that of the assemblyman or woman, one can count on their total support to mobilize the study community. Chiefs in Ghana are highly respected by their subjects and wield a lot of influence over their subjects.

The last stage of the chart is to meet with the people in the community. This usually takes the form of a durbar which is organized by the chiefs and elders in conjunction with the assemblyman or woman. The proposed study is presented at the durbar in a language that the people understand. The chiefs and elders also address the durbar to declare their support for the project and request their subjects to fully participate in the study. Every thing about the study must be explained at the durbar and the people are allowed to ask questions or seek clarifications.

Current and Future Status of Biomedical Research in Ghana

Currently, there are varied research activities ranging from applied health research to various basic health researches and the future is bright for both locally initiated and collaborative health research in the country. However, we need to remind ourselves of the need to reward study participants appropriately but not to take advantage of the people. This is where I believe that we need to take a cue from what is currently being done in the developed countries like the USA in terms of rewarding study participants.

I have observed from various European and American Institutions that, all research advertisements consider subjects time as wasted to participate in the advertised study and therefore, clearly state that all willing and qualified participants that will be selected would be paid for both their time and transport fare. Thus, study participants are given monetary rewards and people are also sometimes motivated with money to keep their appointments. I sincerely believe that there should not be double standards on the way we treat research participants either in rural Africa or in for example metropolitan Baltimore, considering the fact that African respondents are more likely to be poorer than those in the developed world, we need to identify with their plight and also motivate them as it is done in the developed countries. This does not mean paying people to participate but to reward them for their time.

My experience has shown that giving incentive to study participants after they are recruited into a study helps to run the study very smoothly, especially studies that are longitudinal and requires a number of sample taking. This is because human rationality will not allow people to continue to take 'thank you' as rewards for many times and this could affect commitment to the study. In most of my research activities, I do provide a minimum motivation to study participants usually in the form of refreshments. However, I must admit that on several occasions, respondents have asked me to convert it into cash for them and their reason is simple, they cannot leave their family at home, sometimes without fish or meat to eat or soap for washing and be enjoying themselves.

Also, I want to advice that no research team should be involved in local or national politics as this could jeopardize the research work. It may leads to the project being identified as belonging to a particular group of people or supporters of a particular political party and therefore excludes others in the community. Also never identify your project as a government project if it is not one, although, political leaders at the local level will like to hijack your project for it to looks like a government project, be quick to explain to the people the source of your funding and how your work is complementary to what government is doing but not a government project. No government in power has the total support of the people, hence the need to thread on political ground circumspectly. If you do these things, you will find out that Ghanaians are the African's friendliest people and you will enjoy the proverbial Ghanaian hospitality.

Conclusion

It may not to advisable to concentrate research activities in one or a few communities as this could lead to respondents' fatigue. Although, it is very easy to think that once you have established rapport with a particular community all your research activities will be accepted, this could be true only for sometime because the people may be tired of you or being asked similar questions over and over again, especially if your work is not an intervention which brings some immediate personal benefits to the people.

It is important to pay attention to all the processes outlined in this document so as to have a fruitful research carried out in Ghanaian communities without difficulties. Every body from local people through the chiefs and elders to the health officials is very important for the success of the study. Also the institutional clearance is very essential to have some legal backing before entering into a community to conduct health research. Confidentiality and trust is the key here, make sure that you keep all your promises and those that you cannot fulfill due to unforeseen circumstances, go back to explain it to the people and reassure them. Furthermore, there is a need to consider some form of motivation to compensate for respondents' time and energy put into the research. As a respondent once said to me: 'if you do not go, you may not eat' referring to going to collect daily foodstuff from the farm.

Acknowledgements

I would like to thank Prof. D Ofori-Adjei, Director, Prof. M. Wilson, Deputy Director, Prof. K Koram, head Epidemiology department, Dr. Daniel Dodoo all of, NMIMR and Dr. Mrs. Naa Amanuah Chinbuah, Health Research Unit, MoH for their various contributions to the compilation of this chapter in the African Biomedical Research Book Compilation. I am also grateful to Mrs. Cynthia Ahorlu for her support. This contribution will be useful for any person seeking to conduct biomedical or health research in Ghana.

References

Ahorlu, CKS (2005). Cultural epidemiology for malaria control in Ghana. Ph.D. Thesis submitted to the Swiss Tropical Institute, University of Basel, Switzerland

National Bureau of Statistics, Ghana (2002). National Population Census General Report

Republic of Ghana (1998) Policy guidelines for strengthening Research to support the Medium Term Health Strategy in Ghana. Health Research Unit, Accra, Ghana.

In: A Gateway to Biomedical Research in Africa
Editor: Takafira Mduluza, pp. 141-146

ISBN: 1-60021-444-4

Chapter XIV

Concluding Remarks

Towards Community Engagement

Research program entail irrefutable requirement for investment into addressing the complex community challenges of successful trial implementation during the critical era of collecting biological materials, drug development and vaccine development trials. African communities have a lot of respect for health professionals. However the moment some crooked behavior is noticed the trust can easily wane out. There are several studies that have led to failure in clinical trials like the ones on the tenofovir trials in Africa and South America. The community initially agreed to participate but then realized in the conduct that the explanation was not all that clear at the beginning which means there wasn't enough investment in the community engagement. One of the major failures is caused by mistrust or inequitable distribution of worker remunerations or community contact with foreign researchers instead of using local personnel. In some cases due to greedy by some researchers from developed country funding agents need to save and pay very little amounts to local counterparts playing that major role in recruiting participants resulting in exerting minimal effort in that community engagement. Being a foreigner and running the show creates mistrust in the participants, while foreign researchers rarely understand the local norms and beliefs that only a local seasoned researcher would understand and engage the community effectively in local language explaining the study in local understandable terminology. Engaging community stakeholders or members in biomedical research is less of a controlled and predictable format not clearly straightforward. The biomedical trials come a long way and have involved several input in every aspect totaling millions of dollars yet when it comes to the final end user, nothing is invested or used. This brings in the hot and undebatable issue of inputting finance and remunerations.

Incentives: Monetary and other Compensation

Sponsors and even researchers invest millions of dollars in product development, clinical trials, design and building of facilities and employing even none well informed personnel to run or be in charge of the trials but leaving out vital processes of community mobilization. What is hard to believe is the general dominating fact that researchers in developing areas do not need to be awarded competitive remuneration as their counterpart from the funders? At the same time the preclinical trial phases conducted in the sponsor country involve very few participants but the cost is really extra-nomical almost equivalent to the actually late phase III trial that involves several participants. Not understandably, whoever decided that the participant or communities in developing countries do not warrant any remuneration? One prominent researcher explained this attitude to be equivalent to taking advantage of illiteracy and the belief that all health activities ultimately benefit the participants since there is delivery of free services. However, how long will this belief be taken by the African participant? Literacy is now rising in Africa and above everything else the economical changes in several afflicted countries is making participants be aware of their rights and the rational of not to participate. Developed countries have different levels of compensation as indicated below and as in Annex A. Surprisingly, nothing of the similar nature can be cited from any of the biomedical and clinical trials siktes cnducted in Africa or any developing country.

Many leading African researchers feel that monetary compensation should be given to clinical and biomedical participants. Researchers do not realize the amount of time that is wasted participating in biomedical protocols by participants. This has been considered in developed countries where a participant is compensated for time spent in a study. Examples being as presented in Annex A, where the participant is spending several days sleeping in a study which in Africa is never heard of being paid for sleeping? Considering the participant in Africa where it is uneconomical to spend the whole day queuing for blood draws or for questionnaires to be administered. Above all considering that most of the participants willing to be involved are women who should devote their time to taking care of gardens and animals. There is massive loss of man-hour during most biomedical participation in developing countries. But the ingenuity of the sponsors and funders in developing countries take this to be an expensive exercise that should never be discussed at length. The attitude may not continue for long before every corner of the universe realizes the unfair treatment of biomedical research participation. A concern of some misuse of the monetary reward or compensation is alluded to but once one is compensated for time loss during participation the onus is upon the participant to decide on how to spend the award. However, compensation can be in the form of hard cash, goods, foodstuffs, housing or donation to institutes and supporting organization. Sometimes it is unbelievable as to the reason behind denying the participant direct benefit when it is the individuals' time and resources are affected. Channeling the benefits to some institutes creates illegal benefits by the leaders of the institutes since most of the proceeds are not directly accounted for or expected by the regional or country offices. In the long run the habit would create unbearable non compliants in such regions where the participants were told or realized that their participation was compensated for but the benefits were channeled and ultimately misused elsewhere. There is stalk differences between developed country participants and the poor participants who is in the developing regions. It is not on ignorance but twisting of the information to safeguard future biomedical research in case the local indigenous is empowered to benefit from compensation. The attitude was established and broadcasted in fear that most trials would run into bankruptcy if to compensate for the phases where large participants are involved, while this opened the way for free, voluntarily participation. From assesses by the experienced African researcher contributing from the developing countries in this document it is something that will not last long in Africa. Participants are now aware of the double standards of the funders and the researchers both from the developed and developing countries. Denying the poor man the dues supposed to them may create a difficulty attitude to work with in developing countries, as many become aware of the standards.

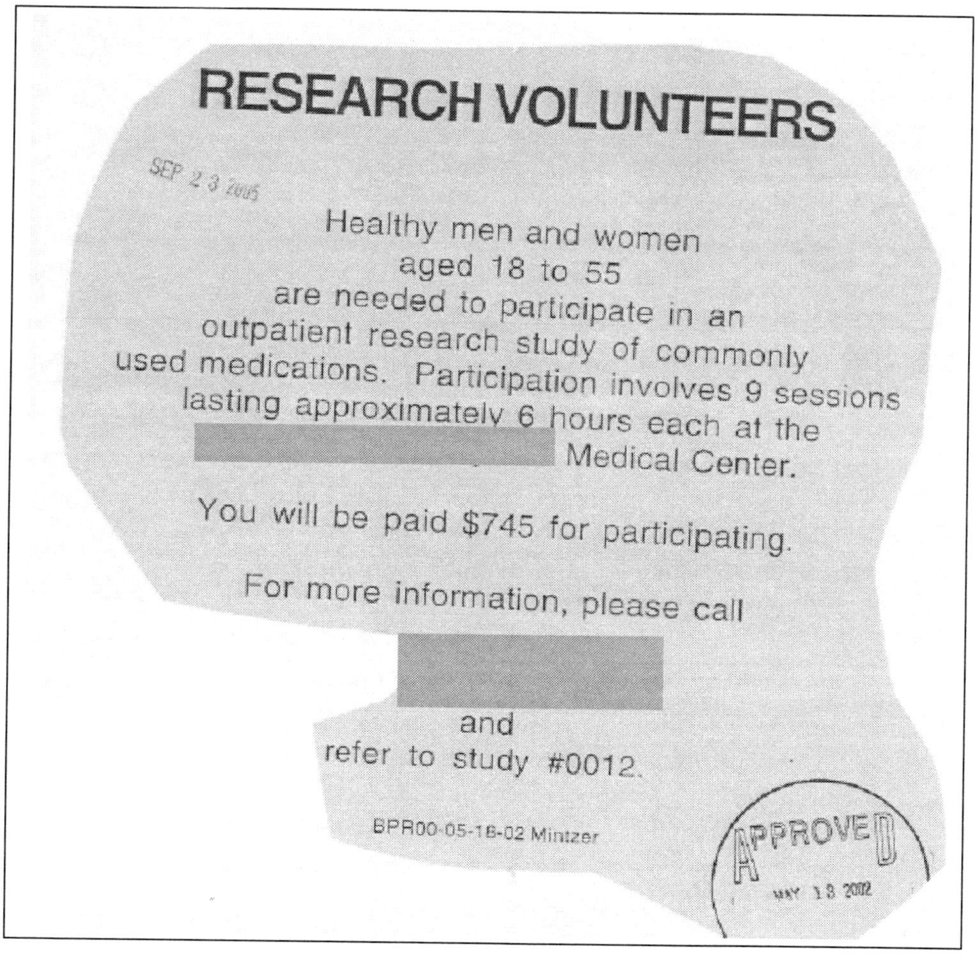

Benefits to participants can be in different ways that should include comprehensive follow-up and provision of treatment. The later has been used for a very long time now in Africa where there are several diseases but at least was considered as a benefit to accessing medications during the trial or research period. However the period after the trials has been neglected. Those who benefited only managed during the program, the benefits were not continued after the surveys. Nothing in reality occur in terms of medical care as well as care relating to illness discovered during the clinical trials. Researchers should make it clear to participants that at some point following the trial the treatment would be of limited duration and would be stopped once the trial is over and as such clearly outline the compensation due instead of hiding behind non compensatory attitude because of costs or due to governments some of which have no obligations to its citizens. Continued care, monetary reimbursements as appropriate compensation for participants helping the investigation to a success. This should be weighed against the benefits the investigators would obtain from the study. But who would do the evaluation on the benefits post study on the investigator? The RECs members should be able to evaluate this and come up with the levels of compensation tallying the estimated dues from such trials. This may assist the researcher to practice and fulfill their obligations to the participants regarding informed consent and also community regarding recruitment, and benefits. The need for adequate informed consent focusing upon therapy duration, risks/benefits and alternatives. The concerns can be summarized as the cardinal issues depending on the type of the biomedical research. There is no need to hoodwink the participants by desperation for therapy, or accepting a risk for knowledge, and in turn expect some form of benefit, which can be true in drug trials of the infectious diseases that have no cure like HIV/AIDS. In as much that knowledge may be required in some current diseases of no cure or vaccine, there is much concern for exploiting the participants, whether by unfairly inducing participation, or by making use of human subjects for reasons unrelated to the health needs of the host country.

Participants in less developed country have a desire for education and are normally eager to contribute to understanding diseases mechanisms. It is the duty of the researchers to clearly give out the information and everything necessary for the execution of the study. Participants view researchers as a source for education. There is a unique relationship between the participant and the researcher in most less developed setting in rural Africa and thence there is a potential for such a relationship to be enhanced or taken advantage on the defenseless and trusting population. There is a need to separate the traditional belief of a researcher or biomedical worker as enhancing good living standards in the community to the investigative researchers whose role is to fulfill research agenda. Safeguarding the ethical norms of the study population is very important and that power has been vested in some regulatory body in the concerned working areas. Currently there is no doubt that we have sufficiently well versed personnel in all ethical norms of protecting and safeguarding the participant but it is the remuneration that is lacking. Empowerment of the guardians of the research ethics seem to spend exorbitant amounts on module after module while it would best invest in strengthening the RECs and have some remuneration. The best option would be to work along the studies seeking participation in less developed countries to foot the bill on both RECs reviewing process and then monitoring and taking into account the remuneration that whatsoever the reason should not be on a different scale from the preceding studies in developing countries and the trial participants. Remuneration is something that researchers cannot hide away from the participants in developing communities in the near future as populations become aware and the research global community becomes a village.

Concerns

- Monetary Compensation - for trial participation and community education and awareness.
- General health care during follow-up and also on illness discovered during the screening stage and trials.
- Moral obligation of the researchers and sponsors to the participant and the community at large.
- Duration of study to be clearly explained in the consent process.
- Local researcher to go a mile longer in ensuring that thorough informed consent is obtained, that the trial is a true representation of all socio-economic strata of the population.
- The researchers to have paternalistic duty to protect the participant.
- Explaining the benefits of the investigator from the study and also of the participant whose role is helping the researcher meet project task and obligation.
- Explain the likely side effects in participating and the likely pain one could experience including latrogenic adverse events if it is drug or vaccine.

International Declaration of Health Rights

The enjoyment of the highest attainable standard of health is one of the fundamental rights of every human being. It is not a privilege for those with power, money or social standing. Health requires a sustainable environment with balanced human and population growth and preservation of cultural diversity. Health depends on the availability to people of basic essentials: food, safe water, housing, education, productive employment, protection from pollution and prevention of social alienation. Health depends on protection from exploitation without distinction of race, religion, and political belief, economic and social condition. Health cares should be based on dialogue and collaboration between citizens, professionals, communities and policy makers. Health services should be affordable, accessible, effective, efficient and convenient. Ultimately, health requires peaceful and equitable development and collaboration of all peoples.

Glossary/Annex

A: Research Participant Informed Consent and Privacy Form

Sites of Research:

Addresses of site of research

Patient I.D. Plate

Research Participant Informed Consent and Privacy Authorization Form

Protocol Title:
Application No.:
Sponsor:
Principal Investigator:
Date:

1. What You Should Know about This Study:

- You are being asked to join a research study.
- This consent form explains the research study and your part in the study.

- Please read it carefully and take as much time as you need.
- Please ask questions at any time about anything you do not understand.
- You are a volunteer. If you do join the study and change your mind later, you may quit at any time without fear of penalty or loss of benefits.
- While you are in this study, the study team will tell you any new information that could affect whether you want to stay in the study.

2. Why is this Research being Done?

This research is being done to evaluate the effects of disrupted sleep patterns and sleep deprivation on the perception of pain, and other physical and psychological symptoms. This study will help researchers understand the relationship between sleep and pain, and how sleep disturbance might influence chronic pain conditions.

Males and females who are: healthy, good sleepers, between the ages of 18 and 45, may join. You have been asked to take part, because you are a non-smoker, without major medical or psychiatric problems. Approximately 100 people are expected to participate in this study.

3. What Will Happen if You Join This Study?

If you agree to be in this study, we will ask you to do the following things:

This study will be conducted in 2 parts:
Part 1 will involve a screening period lasting ....
Part 2 will involve ....
If you agree to be in this study, we would ask you to do the following things:

Part 1: 2 Screening Visits

Before starting the experiment (Part 2) you will complete 2 screening visits to help determine whether you are eligible to participate.

Screening Visit 1: During your first screening visit, you will be asked to complete several **questionnaires** about your medical symptoms. During this visit a small sample of blood will be drawn (3-4 teaspoons) and tested to help us determine that you are healthy and not taking certain medications and drugs that might influence ....

These **laboratory blood tests** will include a standard ....

The other laboratory tests include a test of your liver function, a pregnancy test, and a test of your blood cell counts. At this visit, a nurse will conduct a standard physical exam. This visit will take approximately 2 hours. At the end of this visit, you will be asked to .....

Screening Visit 2: During your second visit, we will review the results of the blood tests and to determine whether you are eligible to participate further. You may be provided a copy of the results upon request. If you continue to remain eligible, you will be scheduled to report to the study clinic centre.

Part 2: Questionnaires

You will be asked to arrive at early, where you will be registered. You will have your blood pressure taken and complete standard questionnaires related to daytime symptoms you may experience, including questions about your general health history. You will complete these same questions each time we visit you or sampling. These questions will take approximately 30-minutes to complete.

Laboratory. You may discontinue any of these procedures at any point simply by telling the experimenter that you wish to stop.

Biological samples

Cognitive Testing

Group 1. If you are assigned to Group 1

Group 2. If you are assigned to Group 2, you will be asked to ......

Group 3. If you are assigned to Group 3, you will be asked to ......

Group 3. If you are assigned to Group 3, you will sleep uninterrupted for an 8-hour period as previously described.

During your participation, any significant or abnormal findings from your medical exam or the laboratory tests of your blood will be discussed with you by the nurse or study physician.

4. What Are The Risks Or Discomforts Of The Study?

Laboratory Testing.

The procedures are likely to produce discomfort, but the intensities used are not enough to cause harm.

Laboratory Blood Testing

The risks of drawing blood include discomfort, bleeding or bruising where the needle enters the body, and in rare cases, fainting or infection.

There may be side effects and discomforts that are not yet known. During this study, you will be told any new facts that could affect whether you want to stay in this study.

Questionnaires.

There is a very minor risk that you may become fatigued or distressed by completing the questionnaires and other tests. Some questions ask about diseases symptoms, which might be considered sensitive.

Sensitive Information.

The laboratory blood testing will identify the presence of infection (if detectable), which might have consequences to you if the results were released to others. Every effort will be made to maintain your privacy and confidentiality.

5. Are There Risks Related to Pregnancy?

Pregnancy is associated with.., so women who are pregnant or become pregnant are excluded. All participants will be required to take a pregnancy test prior to beginning the study. Women found to be pregnant will not be permitted to participate further. This research may hurt an embryo or fetus in ways we do not currently know.

6. Are There Benefits to Being in the Study?

There is no direct benefit to you from being in the study. However, there is a possibility that participation in this study might identify a medical condition, which is currently unknown to you. In this case, this information will be discussed with you and we will help provide you with an appropriate referral for follow up care. Taking part in this study will help increase the scientific knowledge of the diseases, which may help improve the treatment of individuals who have, or are at risk for chronic disease conditions.

7. What Are Your Options if You Do Not Want to Be in the Study?

You do not have to join this study. If you do not join, your care at the clinic will not be affected.

8. Will It Cost You Anything to Be in This Study?

The study procedures will be provided at no cost to you.

9. Will You Be Paid if You Join this Study?

You will receive a total of $900.00 for completing all of the study procedures (two screening visits + seven and a half days of testing and studies). You will be paid after completing the study. If you choose to leave the study early for any reason, you will receive partial compensation as follows: $25 for completing Visit 1, $25.00 for completing Visit 2 (includes completing 2 weeks of diaries). During the experimental phase, you will be paid $50.00 for each night spent at the Research Center and $50.00 each day for completing daytime questionnaires and tests (evening and morning testing). If you complete all of the study procedures (two screening visits + seven and a half days of testing and sleep studies) you will earn an additional $100.00 for your effort.

10. Can You Leave the Study Early?

Your participation is completely voluntary and you may decide to leave for any reason:

1. You can agree to be in the study now and change your mind later.
2. If you wish to stop, please tell us right away.
3. Leaving this study early will not stop you from getting regular medical care.

However, if you participate in this study, you must agree to a) not drive yourself home after the disruption, restriction, or deprivation portions of the study; and b) not drive, operate heavy machinery, or engage in other potentially harmful activities until you have obtained at least 8 hours of sleep.

11. Why Might We Take You Out of the Study Early?

You may be taken out of the study if:

1. Staying in the study would be harmful.
2. You need treatment not allowed in this study.
3. You fail to follow instructions.
4. You become pregnant.
5. The study is cancelled.
6. There may be other reasons that we don't know at this time to take you out of the study.

12. How Will Your Privacy Be Protected?

The Research team has rules to protect information about you. Federal and state laws also protect your privacy. This part of the consent form tells you what information about you may be collected in this study and who might see or use it.

Generally, only people on the research team will know that you are in the research study and will see your information. However, there are a few exceptions that are listed later in this section of the consent form. Unless you give permission or the board that reviews research studies approves it, no one else will be able to see or use your information.

The people working on the study will collect information about you. This includes things learned from the procedures described in this consent form. They may collect other information including your name, address, date of birth, and other details.

The research team will need to see your information. Sometimes other people at Hospital/clinic may see or give out your information. These include people who review the research studies, their staff, lawyers, or other research staff.

People outside of the research team may need to see your information for this study. Examples include government groups, safety monitors, and other hospitals in the study and companies that sponsor the study.

We cannot do this study without your permission to use and give out your information. You do not have to give us this permission. If you do not, then you may not join this study.

We will use and disclose your information only for the purposes stated in the study. The use of your information has no time limit. You can cancel your permission to use and disclose your information at any time by calling the IRB at (000-000-0000) or by sending a letter to:

Institutional Review Board Office

Your cancellation would not affect information already collected in this study.

We try to make sure that everyone who needs to see your information uses it only for the study and keeps it confidential – but we cannot guarantee this.

13. What Other Things Should You Know about This Research Study?

A. What is the Institutional Review Board (IRB) and how does it Protect You?

The IRB is made up of:

- Doctors
- Nurses
- Ethicists
- Non-scientists
- And people from the local community.

The IRB reviews human research studies. It protects the rights and welfare of the people taking part in those studies. You may contact the IRB if you have questions about your rights as a participant or if you think you have not been treated fairly. The IRB office number is 000-000-0000

B. What Do You Do if You Have Questions about the Study?

Call the principal investigator, at 000-000-0000

C. What Should You Do if You Are Injured or Ill as A Result of Being in This Study?

Call Doctor @ (000) 000-0000 or 000-000-0000 (pager) if you have an urgent medical problem related to your taking part in this study.

Call the person in charge of this study, Dr.OOOOOOO, at 000-000-0000, if you think you are injured or ill because of this study.

Medical care at the Clinic/Hospital is open to you as it is to all sick or injured people. The hospital/clinic does not have a program to pay you if you are hurt or have other bad results from being in the study. The costs for any treatment or hospital care would be charged to you or your insurance company.

D. What Happens to Data, Tissue, Blood and Samples That Are Collected in the Study?

Scientists at research institute work to find the causes and cures of disease. The data, tissue, blood and samples collected from you during this study are important to both this study and to future research.

If you join this study:

- The clinic/hospital and/or its outside partners in this research will own these data, tissue and blood samples.
- Scientists may study, test and use this material in this study.
- Scientists may study, test and use this material in future research only with your consent or with the approval of a special review board.
- If this material is used to create a product or idea, the scientists will own that product or idea.
- You will not receive any financial benefit from the creation, use or sale of that product or idea.

E. What Are the Organizations That Are Part of the Research Team?

14. What Does Your Signature on This Consent Form Mean?

Your signature on this form means that:

- you understand the information given to you in this form
- you accept the provisions in the form
- you agree to join the study

You will not give up any legal rights by signing this consent form.

WE WILL GIVE YOU A COPY OF THIS SIGNED AND DATED CONSENT FORM

NOT VALID WITHOUT THE IRB STAMP OF CERTIFICATION

Do not sign after the expiration date of: _______

FOR ADULTS CAPABLE OF GIVING CONSENT:

____ Date
Participant's Signature

SIGNATURE(S):

____ Date
Signature of Person Obtaining Consent
(Investigator or IRB Approved Designee)

____ Date
Witness to Consent Procedures (Optional unless IRB or Sponsor required)

NOTE: A COPY OF THE SIGNED, DATED CONSENT FORM MUST BE KEPT BY THE PRINCIPAL INVESTIGATOR; A COPY MUST BE GIVEN TO THE PARTICIPANT; AND, IF APPROPRIATE A COPY OF THE CONSENT FORM MUST BE PLACED IN THE PARTICIPANT'S MEDICAL RECORD.

FOR OFFICE USE ONLY: **STUDY APPROVED FOR ENROLLMENT OF:** _X_ Adults Only __ Adults and Children __ Children Only

B: Parent Permission (Consent) Form / New Research Project

Title of Research Project:
Principal Investigator:
CHR#

Introduction

This permission form explains the research study and to seek your permission for your child (children) to participate in this study. Please review this form carefully and ask any questions about the study before you consent. You may also ask questions at any time during the study.

Purpose of Research Project

Some children in your community are infected with both malaria and parasites known as Schistosome parasites (which cause bilharzia). This study will determine if there is increased risk to health of your child if infected with both parasites at the same time.

Procedures

- Briefly, this study will follow a group of 500 children over 3 months during two successive malaria seasons. During the study: At the first visit, a questionnaire will be filled out by asking question from you and your child. It should take less than 15 minutes of your time.
- Children will be visited at their schools and will not have to make special trips for follow-up visits. At the first visit we will measure height and weight of your child (children) and also externally examine if liver and spleen are enlarged because of pre-existing infection. In addition, your child (children) will be asked to provide urine and stool samples to determine *schistosoma* infection.
- At the first visit, we will do a finger prick to get drop of blood to test your child for malaria infection. If your child is found to have malaria symptoms (mainly high fever), your child may be treated for malaria in accordance with the Ministry of Health recommendations at this visit. However, prior to treatment, we will also collect blood 10ml (equals 2 teaspoon) from children of 8 years of age or older) and 5ml (from children <8 years age) for other hematological and immunological tests.
- At the next visit 3-5days after the first visit), blood 10ml (equals 2 teaspoon) from children of 8 years of age or older) and 5ml (from children <8 years age) will be collected from all participants for hematological and immunological tests. Based on blood slide examination, if your child is found to have malaria, your child will be treated for malaria in accordance with the Ministry of Health recommendations. At this visit, children also found to be positive for Schistosomiasis will be treated at the local health clinic with the standard drug used for treatment (praziquantel 40mg/kg body weight).
- At 8-12 weeks after treatment for Schistosomiasis, urine and stool examinations will be repeated to make sure of successful treatment. If necessary, treatment will be repeated.
- We will visit your child two more times, six months after the 8-12 weeks visit, and 9-12 months after the 8-12 week visit. At the time of visit, we will record if your child has malaria-like symptoms (such as fever). This may require a finger prick to prepare a blood film as above for examination. If necessary, malaria treatment will be repeated as above. There will be a minor disruption of regular school activities.
- A repeat blood collection in the same amounts may be necessary to successfully complete the study at one of the follow up visits.
- We will also ask that you allow us to get information about your child's health from the school health teacher and the state nurse. We will ask you to sign a separate form to allow us to do this.

Risks/Discomforts

Risk of bruising or infection following a blood draw is a possibility. However, we will take steps to reduce this risk.

Benefits

Children will also receive free medical treatment for malaria and schistosomiasis infections and referred to health centers for other conditions. Children will also be followed up and offered re-treatment, if necessary. Please note that these treatments are also available to you at your local health clinics, although at a nominal fee.

Confidentiality

Participants will be identified by an ID number and their residential address will not be entered on the questionnaire. Although malaria and schistosomiasis are not stigmatized in the study population, any linkage to name will be entered in a secure database accessible only to the PI and study investigators. Records will be kept in a secure locked cabinet in an office of the PI at Blair Research Laboratory in Harare. Other records will be kept on a password-protected computer of the research investigators. Identifiers or any links/codes to any identifiable information that would make it possible to identify you or your child will be removed after 4 years.

Compensation

All participants who are recruited and are available for all the visits (initial consent, initial sampling and the three follow-up visits) will receive an insecticide-treated mosquito bed net to help protect them against future malaria mosquito exposure.

Voluntariness

Your child's participation in this research project is completely voluntary. You have the right to refuse your child's participation in this study. You have the right to withdraw your consent from the research study at any time. Even if you do not want to join the study, or if you withdraw from the study, you will still receive the same quality of medical care available to you at health clinics. You should ask the principal investigator listed below any questions you may have about this research study. You may ask him/her questions in the future if you do not understand something that is being done. The investigators *(or doctors)* will share with you any new findings that may develop while you are participating in this study.

Persons to Contact

The purpose of the study and the procedures has been explained to you for your child to be in the study. However, if you want to talk to anyone about this research study because you think you have not been treated fairly or think you have been hurt by joining the study, or you have any other questions about the study, you should call the Local principal investigator Dr. Takafira Mduluza, at 263-4-303221 (extn. 1344) or call the Medical Research Council of Zimbabwe, P.O. Box CY 573, Causeway, Harare (263-791792). The principal investigator or the people in the Medical research Council will answer your questions.

If you have read this document and you have been given the chance to ask any questions now or at a later time or if the document has been read and explained to you and you agree for your child to be in this study, please sign or make your mark below.

Print Name of Subject: ____

____ ________________

Signature or Mark of Subject or Legally Authorized Representative — Date

____ ________________

Signature of Person Obtaining Consent — Date

____ ________________

Witness to Consent if Subject Unable to Read or Write Date
(Must be different than the person obtaining consent)

Signed copies of this consent form must be 1) retained on file by the principal investigator, 2) given to the subject and 3) placed in the subject's medical record (when applicable).

NOT VALID WITHOUT THE CHR
STAMP OF APPROVAL

CHR#: ________________________
VALID FROM____________ TO______________

C. I: ASSENT DOCUMENT (Children 7 To 10 Years Age)

Project Title: Malaria and Schistosoma co-infection in Zimbabwe.

Investigators: Dr. Takafira Mduluza and Dr. Susan Mutambu.

We want to find out if you suffer more from malaria when you are also infected with worms that cause schistosomiasis (also known as bilharzia where you live). If you decide to take part in this study

- We will ask a number of questions in the presence of your parents or legal guardians. It may take about 15 minutes to answer various questions.
- You will be asked to give stool, urine and blood samples during each visit. These samples will determine if you are infected with Schistosome worms and malaria parasites. These sample collections will be at your school. We will visit you at your school for follow up studies (total of 3 follow up visits). You may feel minor pain and/or discomfort, similar to when your blood is taken at other health centers.

When we complete this study we will write a report. This report will not include your name, or indicate that you were in this study. The study will suggest better ways to treat these infections in your community and protect you in future. You will receive standard treatment (recommended by your Ministry of Health) for Schistosomiasis and malaria. You will also get some education about these health problems and how to protect yourself in future.

You do not have to be in this study if you decide that you do not want to. Your participation is absolutely on voluntary basis. If you decide to stop participating after we begin, you can inform us directly or through your parents and that will be OK too.

If you decide you want to be in this study, please sign your name.

-- ------------

(Sign your name here) (Date)

C. II: Assent Document (Children 11 To 17 Years Age)

Project Title: Malaria and Schistosoma co-infection in Zimbabwe

Investigators: Dr. Takafira Mduluza and Dr. Susan Mutambu.

We are conducting a study to determine if you are at increased risk for malaria infection when you are also infected with worms that cause schistosomiasis (also known as bilharzia where you live). If you decide that you want to take part in this study, you will be asked to answer a number of questions in the presence of your parents or legal guardians. It may take about 15 minutes to answer various questions.

In addition, you will be asked to provide stool, urine and blood samples during each visit to determine if you are infected with Schistosoma and malaria parasites. These sample collections and future visits will be at your school. We will visit you at your school for follow up studies (total of 3 follow up visits). You may feel minor pain and/or discomfort, similar to when your blood is taken at other health centers.

These studies will suggest improved ways to treat these common infections in your community. During your participation, you will receive standard treatment (recommended by the Ministry of Health) for Schistosomiasis and malaria. Moreover, you will receive education on these health problems and how to protect yourself and others in your community. When we are finished with this study we will write a report. This report will not include your name, or indicate that you were in this study.

You do not have to be in this study if you decide that you do not want to. Your participation is absolutely on voluntary basis. If you decide to stop participating after we begin, you can inform us directly or through your parents and that will be OK too.

If you decide you want to be in this study, please sign your name.

--- ------------------

(Sign your name here) (Date)

D: Script for Study Information

Example:

Study Title: Malaria and Schistosoma co-infection in Zimbabwe

Study Investigators: Dr. Takafira Mduluza (University of Zimbabwe, Harare), Dr. Susan Mutambu (National Institute of Health Research, formerly Blair Research laboratory, Harare).

- Some children in your community are infected with malaria parasites and parasites known as Schistosome parasites (which cause bilharzia). These parasites pose serious risk to health of your child.
- This study will determine if there is increased risk to health of your child if they are infected with both parasites at the same time.
- We will ask a number of health related questions in the presence of parents and legal guardian.
- We will need to examine urine and stool for infection with parasites causing bilharzia (schistosomiasis). If found infected, we will treat with drugs to clear infection and make sure that the infection is completely cured.
- We will examine blood films to detect malaria infection. If infected, we will treat for malaria infection. We will also carry out other tests on blood to look for other malaria disease related changes.
- During follow up visits, we will carry out similar blood examinations.
- Results of this study will suggest methods to minimize risk to health of your child when infected with these parasites.
- Please note that your child's participation in this research project is completely voluntary. You have the right to refuse your child's participation in this study. You have the right to withdraw your consent from the research study at any time. Even if you do not want to join the study, or if you withdraw from the study, you will still receive the same quality of medical care available to you at health clinics.

E: Certification of Translation

COMMITTEE ON HUMAN RESEARCH

CHR #: ____________________________________

Project Title:

I, ____, certify that, to the best of my knowledge, the translation of the consent document(s) from English to the local language of________________________________, for the project indicated above is accurate.

____ ____________________

Signature of Person Providing Certification Date

____ ____________________

Organization/Institution Title

____ ____________________

Signature of Principal Investigator Date

NOTE: The Investigator(s) on the study may not certify the translation; however, the Principal Investigator=s Co-Signature is Required.

F: Appendices for Biomedical Research in Zambia

Appendix 1: (a)The University of Zambia Research Ethics Committee (UNZA-REC) Research

Proposal Requirements

THE UNIVERSITY OF ZAMBIA

Research Ethics Committee

Telephone: 260-1-256067
Telegrams: UNZA, LUSAKA
Telex: UNZALU ZA 44370
Fax: + 260-1-250753
E-mail: unzarec@zamtel.zm

Ridgeway Campus
P.O. Box 50110
Lusaka, Zambia

Assurance No. FWA00000338

IRB00001131 of IOR G0000774

Information Required for Submitting a Research Proposal

1. Documentation

For a thorough and complete review, all research proposals shall be submitted with the following documents:

i. Name of the applicant with designation.
ii. Name of the Institute/Hospital/Field area where research will be conducted.
iii. Approval of the Head of the Department/Institution
iv. Protocol of the proposed research as prescribed by UNZAREC format.
v. Ethical issues in the study and plans to address these issues.
vi. Proposal should be submitted with all relevant enclosures like proformae, case report forms, questionnaires, follow-up cards, etc.

vii. Informed consent process, including patient information sheet and informed consent form in local language.

viii. For any drug/device trial, all relevant pre-clinical animal data and clinical trial data from other centres within the country or outside as the case may be.

ix. Curriculum vitae of all the investigators with relevant publications in last five years.

x. Any regulatory clearance required.

xi. A detailed budget and source of funding.

xii. Other financial issues including those related to insurance.

xiii. An agreement to report all Adverse Events promptly to UNZAREC.

xiv. Statement of any conflict of interest.

xv. Agreement to comply with the relevant national and applicable international guidelines.

xvi. A statement describing any compensation for study participation (including expenses and access to medical care) to be given to research participants; a description of the arrangements for indemnity, if applicable (in study-related injuries); a description of the arrangements for insurance coverage for research participants, if applicable; all significant previous decisions (e.g. those leading to a negative decision or modified protocol) by other ECs or regulatory authorities for the proposed study (whether in the same location or elsewhere) and an indication of the modification(s) to the protocol made on that account. The reasons for negative decisions should be provided.

xvii. Plans for publication of results – positive or negative – while maintaining the privacy and confidentiality of the study participants.

xviii. Any other information relevant to the study.

2. Scientific Design and Conduct of the Study

i. Adequate background information and literature review.

ii. The appropriateness of the study design in relation to the objectives of the study, the methodology, statistical analysis including sample\size calculation and the potential for reaching sound conclusions with the smallest number of research participants.

iii. The justification of predictable risks and inconveniences weighed against the anticipated benefits for the research participants and the concerned communities.

iv. The justification for the use of control arms.

v. Criteria for prematurely withdrawing research participants.

vi. Criteria for suspending or terminating the research as a whole.

vii. The adequacy of provisions made for monitoring and auditing the conduct of the research including the constitution of the Data Safety Monitoring Board (DSMB).

viii. The adequacy of the site including the supporting staff, available facilities and emergency procedures.

ix. The manner in which the results of the research will be reported and published.

3. Recruitment of Research Participants

i. The characteristic of the population from which the research participants will be drawn (including gender, age, literacy, culture, economic status, and ethnicity).
ii. The means by which initial contact and recruitment is to be conducted.
iii. The means by which full information is to be conveyed to potential research participants or their representatives.
iv. Inclusion criteria for research participants.
v. Exclusion criteria for research participants.

4. Care and Protection of Research Participants

i. The competence of the investigator(s) (qualifications, experience etc) to carry out the proposed study;
ii. Any plans to withdrew or withhold standard therapies for the purpose of the research and the justification for such action;
iii. The medical care to be provided to research participants during and after the course of the research.
iv. The adequacy of medical supervision and psycho-social support of the research participants.
v. Steps to be taken if research participants voluntarily withdraw during the course of the research.
vi. The criteria for extended access to, the emergency use of, and/or the compassionate use of study products.
vii. The arrangement, if appropriate for informing the research participants, general practitioner (family doctor) including procedures for seeking the participant's consent to do so.
viii. A description of any plans to make the study product available to the research participants following the research.
ix. A description of any financial cost to the research participant.
x. The rewards and compensations for research participants (including money, services and/or gifts).
xi. The provisions for compensation/treatment in the case of injury/disability/death of a research participant attributable to participation in the research.
xii. The insurance and indemnity arrangements, if required.

5. Protection of Research Participants Confidentiality

i. The people who will have access to personal data of the research participants, including medical records and biological samples; and
ii. The measures taken to ensure the confidentiality and security of personal information concerning research participants.

6. Biological Specimen

i. A full description of any specimen that will be collected (blood, body fluids, tissue biopsies etc;
ii. Plans for obtaining consent and clearance from participants and UNZAREC for long-term storage, to export, and future research; and
iii. Arrangement for disposal.

7. Informed Consent Process

i. A full description of the process for obtaining informed consent, including the identification of those responsible for obtaining consent;
ii. The adequacy, completeness and understandability of written and oral information to be given to the research participants, and, when appropriate, their legally acceptable representative(s);
iii. Clear justification for the intention to include in the research individuals who cannot consent and a full account of the arrangements for obtaining consent or authorisation for the participation of such individuals;
iv. Assurance that research participants will receive information that becomes available during the course of the research relevant to their participation (including their rights, safety and well-being); and
v. The provision made for receiving and responding to queries and complaints from the research participants or their representatives during the course of a research project.

8. Community Considerations

i. The impact and relevance of the research on the local community from which the research participants are drawn and the wider concerned communities and the environment;
ii. The steps taken to consult with the concerned communities during the course of designing the research;
iii. The influence of the community on the consent of the individuals;
iv. Proposed community consultation during the course of the research;
v. The extent to which the research contributes to capacity building such as the enhancement of local health care, research, and the ability to respond to public health needs;
vi. A description of the availability and affordability of any successful study product to the concerned communities following the research;
vii. The manner in which the results of the research will be made available to the research participants and the concerned communities.

9. Vulnerable Population

i. Documentation on how the researcher will protect the rights and welfare of categories of population i.e. children, pregnant women, refugees, prisoners, elderly persons, orphans

Biomedical Research in Zambia

Appendix 1: (B). The University of Zambia Research Ethics Committee (UNZAREC)

Proposal Submission Application form

UNZAREC Form No. 1

THE UNIVERSITY OF ZAMBIA

Research Ethics Committee

Telephone: 260-1-256067
Telegrams: UNZA, LUSAKA
Telex: UNZALU ZA 44370
Fax: + 260-1-250753
E-mail: unzarec@zamtel.zm

Ridgeway Campus
P.O. Box 50110
Lusaka, Zambia

Assurance No. FWA00000338

IRB00001131 of IORG0000774

Application for Ethical Approval for Proposed Research Involving Human Participants

To be submitted in 25 copies to the Secretary of the Research Ethics Committee

1. TITLE OF STUDY:

2. PRINCIPAL INVESTIGATOR:

Name: Qualifications:
Present Appointment/Affiliations:

3a. OTHER INVESTIGATORS:
Name: Qualifications:
Present Appointment/Affiliations:
Name: Qualifications:
Present Appointment/Affiliations:
(Other names to be included on a separate page.)

3b. SUPERVISORS:
Name: Qualifications:
Present Appointment/Affiliations:
Name: Qualifications:
Present Appointment/Affiliations:

---

4. SUMMARY OF PROPOSED RESEARCH:

To include aims and objectives, participants to be studied, research methods (questionnaires, physical examination, specimens to be collected, laboratory investigations, standard and experimental therapies, environmental changes, etc.) and statistical analyses. Simple or lay terminology should be used as much as possible.

(Use not more than one additional A4 sheet if necessary.)

---

5. ARE THE PARTICIPANTS DEPENDENT ON ANY OF THE INVESTIGATORS?

As students: Yes/No* As employees: Yes/No*
As patients: Yes/No* In other ways: Yes/No*

If 'Yes" to any of the above, give details:

---

6. POSSIBLE BENEFIT TO PARTICIPANTS:

---

7. POSSIBLE RISKS TO PARTICIPANTS:

---

8. POSSIBLE BENEFITS TO THE COMMUNITY:

9. BUDGET:
(a) Financial support (requested or granted): Yes/No* **Sponsor:** Yes/No*

(b) Are there costs which will be carried by other institutions (e.g. the Hospital)? Yes/No*

(c) Are there costs which will be carried by the participants involved (e.g. travel, accommodation, meals, treatment)? Yes/No*

(d) Will the care or the time spent in hospital be prolonged? Yes/No*

If 'Yes" to any of the above, give details:

---

10. SUBMISSION:

Attachments include the following in 4 copies each:

(a) The full protocol Yes/No*

(b) Forms of Questionnaire Yes/No/NA*

(c) Informed Consent Form Yes/No*

(d) Approval from the appropriate Research Committee Yes/No*

---

* Delete as appropriate. NA: Not applicable

---

11. DECLARATION:

I .... (Full Name)

Apply to the Research Ethics Committee of the University of Zambia for approval of the above research proposal involving human participants, as conforming with recognized ethical standards and as not impinging on the rights of the individuals.

Date: .... Signed: ....

PRINCIPAL INVESTIGATOR

Contact Address: ....

....

....

Telephone No.: .... Fax No: .... Email: ....

Full name of Head of Department:

....

Signed: .... Date: ....

HEAD OF DEPARTMENT

Full name of Supervisor: ....

Signed: .... Date: ....

SUPERVISOR

Appendix 2: The University of Zambia Research Ethics Committee (UNZA-REC) Project

Progress Report Format

UNZAREC Form No. 2

THE UNIVERSITY OF ZAMBIA

Research Ethics Committee

Telephone: 260-1-256067
Telegrams: UNZA, LUSAKA
Telex: UNZALU ZA 44370
Fax: + 260-1-250753
E-mail: unzarec@zamtel.zm

Ridgeway Campus
P.O. Box 50110
Lusaka, Zambia

Assurance No. FWA00000338

IRB00001131 of IOR G0000774

Progress Report Form

The Research Ethics Committee (REC) would like to know how your study has progressed so far, and if any difficulties have been experienced. Please complete the questionnaire below and return, within two weeks, to:-

The Secretary
Research Ethics Committee
Ridgeway Campus
P.O. Box 50110
LUSAKA

1. Name of Principal Investigator:

2. Title of study:

3. REC reference number:

4. Date of REC approval:

5. Briefly describe the purpose of the study (2 – 3 sentences in non-technical language)

6. Has the study started? Yes/No Starting date of study:

7. Number of local research sites recruited: Proposed: Actual:

8. Number of participants recruited into study:

	Male	**Female**	**Age range**
Proposed			
Actual			

9. Number of participants completing study: Actual:

10. Number of withdrawals: Actual:
Reasons for withdrawal:

11. Have there been any difficulties in recruiting participants to the study? Yes/No
If yes, please give details:

12. Have there been any adverse events? Yes/No
If yes, have these been notified to the Committee? Yes/No
Please give details

13. Have there been any amendments to the study? Yes/No
If yes, have these been notified to the Committee? Yes/No
Please give details

14. Has the study been completed? Yes/No Date of completion:
If no, what is the expected completion date?
If the study will not be completed, please give reason(s)

15. Results – please include details of outcomes and conclusions so far (attach a separate page if necessary)

16. Have the findings been disseminated? Yes/No
If yes, how?

Please give details of any publications and send copies when available

17. Any complaints about the research?

Signature of Principal Investigator: ……………………………………………..

Print name: ……………………………………………………………………..

Postal address: ……………………… Tel. No: …………………….. E-mail: …………….

Date of submission:……………………………………………………………..

Index

A

B

C

D

G

H

I

N

O

P

Q

R

V

W

X

Y

Z